Human-Centered Data Science

T0200608

Human-Centered Data Science

An Introduction

Cecilia Aragon, Shion Guha, Marina Kogan, Michael Muller, and Gina Neff

The MIT Press
Cambridge, Massachusetts
London, England

© 2022 Cecilia Aragon, Shion Guha, Marina Kogan, Michael Muller, and Gina Neff

All rights reserved. No part of this book may be reproduced in any form by any electronic or mechanical means (including photocopying, recording, or information storage and retrieval) without permission in writing from the publisher.

The MIT Press would like to thank the anonymous peer reviewers who provided comments on drafts of this book. The generous work of academic experts is essential for establishing the authority and quality of our publications. We acknowledge with gratitude the contributions of these otherwise uncredited readers.

This book was set in Times New Roman by Westchester Publishing Services. Printed and bound in the United States of America.

Library of Congress Cataloging-in-Publication Data

Names: Aragon, Cecilia Rodriguez, author. | Guha, Shion, author. | Kogan,
    Marina, author. | Muller, Michael (Data scientist), author. | Neff, Gina,
    1971– author.
Title: Human-centered data science : an introduction / Cecilia Aragon,
    Shion Guha, Marina Kogan, Michael Muller, and Gina Neff.
Description: Cambridge, Massachusetts : The MIT Press, [2022] | Includes
    bibliographical references and index.
Identifiers: LCCN 2021019747 | ISBN 9780262543217 (paperback)
Subjects: LCSH: Big data—Moral and ethical aspects. | Social
    indicators—Data processing. | Quantitative research—Moral and ethical
    aspects. | Information science.
Classification: LCC QA76.9.B45 A73 2022 | DDC 005.7—dc23
LC record available at https://lccn.loc.gov/2021019747

10 9 8 7 6 5 4 3 2 1

To our students, both past and future, who work to make data science better

Contents

Acknowledgments

The authors wish to thank the many people who contributed to this book, beginning with our patient and highly responsive editor at the MIT Press, Gita Devi Manaktala.

Significant thanks are due to the attendees and co-organizers of the Computer-Supported Cooperative Work and Social Computing (CSCW) 2016 workshop on Human-Centered Data Science, where it first became clear we needed to document this new field. Next, we thank participants in the CHI 2019 workshop on Human-Centered Study of Data Science Work Practices; the GROUP 2020 workshop on Mapping out Human-Centered Data Science: Methods, Approaches, and Best Practices; and the CSCW 2020 workshop on Interrogating Data Science.

We particularly wish to thank Faith Bosworth from Book Sprints for keeping us on track with a gentle but relentless hand throughout our writing workshop in Seattle in February 2020. We also owe Jane Skau and Adeline Swires a great deal of gratitude for providing excellent logistical support during our writing workshop, keeping us well supplied and fed during a period of intense collaborative work.

We wish to thank the Gordon and Betty Moore and Alfred P. Sloan Foundations for their generous support of the eScience Institute at the University of Washington and for funding to help support this work. Support of a British Academy Mid-Career Fellowship is gratefully acknowledged.

We are grateful to Morgan Vigil-Hayes for her recommendations about Indigenous topics, to Theresa Jean Tanenbaum for consultations on gender identity, and to Andrea Simenstad for a careful reading of the final draft.

We thank the anonymous MIT Press reviewers whose detailed and thoughtful comments greatly improved the book and a brilliant team of editors and copyeditors who helped us get it in shape for publication.

Finally, we thank the authors of the case studies for sharing their excellent practical examples showcasing the wide variety in the field of human-centered data science.

1

Data Science to Human-Centered Data Science

On October 9, 2017, Frank Lantz released a simple game based on the following premise: What if we created an AI (artificial intelligence) with an apparently innocent goal: make as many paperclips as possible, as efficiently as possible?

It sounds nonthreatening, even boring. But the clicker game, Universal Paperclips, promptly went viral. And as this story illustrates, the unintended consequences of a relatively simple algorithm can lead to the destruction of the universe. The problem is that an algorithm will execute exactly what its designer told it to do. If the designer forgot to program in bounds or stopping conditions (or common sense, ethics, or human values), the program will continue beyond the designer's original intent for an unbounded amount of time.

In this example, the lack of a stopping condition meant turning all the matter in the universe into 3×10^{55} paperclips.

At a moment of great optimism and enthusiasm for data science, coupled with a rising awareness of systemic societal injustices, this story resonates. Fields as diverse as computer science, artificial intelligence, social science, and even science fiction have been wrestling for decades with similar types of questions about ethics and boundaries and designers' responsibilities. Now we have access to much more data than humans can reliably make sense of, and these kinds of doomsday scenarios, along with algorithmic biases on smaller scales, have become greater risks.

It is common to hear people saying that we can guard against bias or racism in, say, mortgage lending with software that makes a less "biased" evaluation of applicants. That way, you remove "human error" and human prejudice from the equation, right? Those biased bankers or racist lenders will no longer be able to discriminate against people because of the color of their skin or what they were wearing on the day they walked into the office to apply.

In reality, of course, as numerous examples show, algorithms reflect the choices made by their human developers, including conscious and unconscious biases. What's worse, algorithms may amplify these biases, make them less transparent to other people, or make it harder to mitigate them. This is how we get Google Images returning pictures of white men when a person types in "doctor" or highly sexualized results for the phrase "Black girls" (Noble 2018; Wible 2016; see also Bradley et al. 2015). The CEO of a facial recognition software company said in 2018 that the software shouldn't be used by law enforcement to detect criminals because of the inherent racial bias (Brackeen 2018). IBM's CEO went further and declared that IBM was no longer in the facial recognition software business (Allyn 2020; Denham 2020). In June 2020, Microsoft and Amazon followed suit, announcing that they were putting in place a moratorium on selling their facial recognition software to police departments.

Recognition of the problem of *algorithmic bias* is becoming more widespread. The biases of individual designers, which may include the unconscious beliefs of a society's majority population, are inevitably reflected in the design of algorithms. Rather than the "wisdom of the crowd," perhaps we should be talking about the "bias of the crowd."

Beyond reflecting designer bias, the decisions that data scientists encode in algorithms may have unintended individual and societal consequences. Data science has developed significant power over human lives. As data scientists, we must think carefully about that power, its potential consequences, and our responsibilities.

Part of the work in human-centered data science lies in understanding and making transparent the mental models that govern the design of algorithms running over very large datasets. Research in visual analytics and human-centered data science shows that one of the most important elements for maximizing the effectiveness of an algorithm that is designed for humans to use is transparency or understandability (Baldassarre 2016; Brooks et al. 2013; Ye 2013).

And yet, consider the example of deep learning, which is so effective in solving many big data problems but which also creates models so complex, with many variables, that even the system designer cannot always know how a particular output is generated. All we know is that it works. How can we provide an understanding to the user when we don't even know why a model does what it does?

This is a good question, and one that machine learning developers have been wrestling with for many years. In the early 1980s, one of our colleagues attended a machine learning conference and listened to a speaker, a software developer at a bank, discussing the application of a backpropagation network, an early precursor of today's deep learning algorithms. The speaker explained improvements that the machine learning algorithm had made in the bank's mortgage lending process.

The colleague raised his hand and asked, "What do you do when an applicant wants to know why they've been denied a loan?"

The speaker responded, "Because that's a regulatory issue, we are required by law to let the applicant know the reason for any denials. So, what we do is gradually change one of the inputs until we run into the decision boundary of the algorithm. For example, we might raise their income, or increase the number of years they've spent in their job. As soon as the output changes, we can inform the applicant, 'If your salary was X dollars higher, or if your credit rating was so-and-so many points better, you would have received the loan.'"

This story is telling on many levels. First, it illustrates the importance of legal and policy guidelines to govern data and to govern what people are allowed to do with the results of the analysis of data. Second, it demonstrates how legal guidelines may inadvertently encourage developers to reverse-engineer obscure code to satisfy human needs. Even if the user or designer of this algorithm didn't really understand why it produced the results it did, the type of exploration described by this speaker can help produce a useful mental model for the end user. Finally, it speaks to the need to design machine learning algorithms whose parameters are more transparent, or to figure out ways to help people understand what's going on inside the "black box" of the algorithm—which is the goal of the field that has become known as *interpretable machine learning* (Molnar 2020).

Emergence of Human-Centeredness in Data Science

Large-scale data analysis presents opportunities in a wide variety of social, scientific, and technological areas and has led to a variety of innovations (Gluesing, Riopelle, and Danowski 2014; Luczak-Roesch et al. 2015). Yet, increasingly focusing on purely statistical or computational approaches may fail to capture social nuances, affective relationships, or ethical, value-driven community commitments and other human-centered concerns. Human-centered data science emerged to address precisely these issues.

Human-centered data science is a new interdisciplinary field of study that draws from human-computer interaction, social science, statistics, and computational techniques. It draws on the well-established traditions of human-centered design to inform better data science practice. Human-centered data science pushes computational approaches to large-scale data to include the kind of rich detail, contextual knowledge, and deep understanding that qualitative research and mixed methods can bring to the understanding of data and society. The authors of this book have helped to originate this area through writing about these practices, hosting workshops and tutorials to develop them with other researchers, putting them into practice in our own projects, and teaching them in our classrooms. We argue that a deep understanding of the human and social contexts of data helps data scientists to respond to the ethical thinking and choices that they face in their everyday practice. Human-centered approaches also help data scientists develop empathy for the subjects of their data, which can lead to greater acceptance and use of data science systems. Human-centered data science seeks to develop awareness of the complex nature of the interaction between society, technology, and human-generated data. Collectively, we have spent decades working in computational data science and in studying data science. We have come to the inescapable conclusion that without a human-centered approach, data science may not only cause irreparable harm to society but will also be ultimately unsuccessful in its lofty goals. These goals include addressing many widely accepted scientific and societal grand challenges, such as developing new medicines and vaccines trusted by the public, ameliorating hunger and poverty, and creating defenses against climate change.

Human-centered data science is necessarily broad. Our use of the term does not imply that every project should use every element that we define as human-centered. For example, a human-centered approach to crowdsourced scientific data might focus primarily on the dynamics of how people produce the data.

However, human impacts are more pervasive within data science than commonly believed. Often, large datasets for machine learning training are not thought of as being "human-centered." Our approach to human-centered data science, however, suggests making visible the human work that happens, for example, on platforms like Amazon Mechanical Turk so that we can label these training sets. The tools and techniques that we teach in this book help people learning data science to raise these questions, reflect on their work, do their work responsibly, carefully, and ethically, and above all keep the people affected by their projects at the forefront of the planning and execution of their projects.

Our approach goes beyond the "usability" engineering of data science models and pipelines. Our approach is pragmatic and directed at people who do data science projects. We take on board the concerns of social science and humanistic critiques, such as critical data studies. We aim, though, to translate them to an audience of soon-to-be practitioners. In this sense, what you won't find in this book are debates about the theories behind why we do data science the way we do. We are less interested in what we see as a false debate in data science between human-centered approaches and rigor. We show how you as a data science practitioner can build rigorous *and* ethical algorithms, design projects that use cutting-edge computational tools *and* address social concerns, and present results that could speak to statisticians *and* journalists. For us, the human-centered approaches that we present in this book, taken together, create the possibility for people working in data science to have the capacity to make positive changes in the world.

With this book we want to teach you how to reflect on the enormous responsibility that comes with the power of data science. This is because human-centered data science is about the people we study, the people who are involved at practically every stage of the data science cycle, and the people who base decisions on our projects. We show throughout this textbook that there are points where you get to make choices, and this has implications for the power and exclusion of people from this practice. Human-centered data science confounds the notion that automated or computational approaches can be "free" of bias and shows how the power to define categories and ask questions has huge implications for the work that we do.

Doing Human-Centered Data Science

There are multiple ways to "do" human-centered data science. Software developers and data scientists, such as some of the authors of this book, may apply human-centered techniques such as qualitative research to develop better algorithms for data science projects.

A human-centered approach to data science can produce more effective algorithms. One of us (Cecilia) worked as a data scientist and software developer with a group of astrophysicists who had, in their words, "too much" data. They were suddenly facing an onslaught of more than 100 times the amount of data they were used to. The problem was how to tell the good data from the bad. They were analyzing pictures of supernovae, and wanted to automate one of the steps in classifying the images. The team created a requirements document that suggested complete algorithm automation of the process. However, multiple attempts to accomplish this failed. It turned out that the requirements document did not capture the team's underlying needs. By spending time with the team, following human-centered methods that involved careful listening and understanding what they were trying to accomplish rather than merely what they thought they needed, Cecilia proposed new algorithms that were not mentioned in the requirements document but could help the team achieve their goals. For example, solely based on close observation of the team's process, she developed a Fourier contour descriptor algorithm to parse the specific curvature of supernovae from other images in the dataset. This approach turned out to be more efficient and ended up producing 40 percent fewer false positives, significantly improving the team's supernova search process.

Case Study 1.1
Different Ways of Seeing in Data Science
Steven Jackson, Cornell University

I work in *critical* and *interpretive* traditions that study order, value, and meaning as defining attributes of human activity in the world. To do this work, I mostly use ethnographic methods, but I inform them with readings from sociology, anthropology, law, policy, design, and science and technology studies (STS). Now that data science has become so important, I use these tools to study the work of people in data science. In this case study, I am particularly interested in the ethics of data science work.

The "Fairness, Accountability, and Transparency" (FAccT) field has made many important strides in opening the field of data science and algorithmic technique to being studied in terms of ethical assumptions, values, and practices. In addition to these formal studies, there are other virtues and practices essential to the real-world work of data science that are no less important and, taken collectively, constitute the *everyday practical ethics* of the field.

Like all forms of knowledge, data science (whether human-centered or otherwise) provides a *way of seeing*—that is, imagining or picturing the world that inevitably focuses attention on some things and ignores others. Sociologists might describe this as a kind of "standpoint epistemology"—from epistemology (how you know things) and from standpoint (you "see from where you stand"). We learn our standpoints over time: to be a data scientist is to learn to "see" the world in particular ways through the numbers and algorithms of data science (Passi and Jackson 2017). This is a powerful and, in its place, positive development.

Someone who acquires this knowledge can lose sight of the fact that it is one among many ways of seeing. At its worst, this tendency can make data science solutions appear to be inevitable and "objective," and can make data science practitioners seem to have exaggerated authority. A part of this effect may be traced to the stories we (and the world) tell about the nature of work in our field and the cleaned-up stories of practice that ignore the mundane realities of data science work. Despite elevated claims about data science, much of the work is in fact custodial or even janitorial in nature—the incessant effort to gather, clean, and repair data and to make datasets work well enough for the purpose at hand. Rather than a smooth or neutral mirror of reality, data is therefore best viewed as a messy and very human *accomplishment*—the end (or middle) point of a whole world of very ordinary human work.

A different kind of distortion in seeing confronts the attributions of certainty and authority that are sometimes pressed on data science from the outside world. This is a tension that is long familiar to sociologists of science, and this tension is part of the inevitable loss of information during translation from one field to another. A sociologist might call this problem the "reification" of knowledge claims as they move from the researchers who initially generated them (for whom limitations, doubts, and uncertainties are hard to ignore), to more distant users and audiences. For this latter group (especially those with little working knowledge of the real-world practices of data science), the claim or finding may begin to take on a certainty and solidity that will begin to appear magical. In the words of sociologist Harry Collins (1985), "distance lends enchantment"—giving the output of data science work an authority that it may or may not deserve. This may also be seductive and therefore dangerous to data scientists themselves. Who doesn't like to be believed? The sense of authority may also work against the sense of fallibility or "this-might-be-wrongness" that any human-centered data science must carry in its fundamental set of assumptions and work practices.

Finally, data scientists may struggle to recognize the essentially collaborative nature of their work—including with actors who may be closer to the everyday work of the domains or problem areas they seek to address and therefore have powerful standpoints of their own to contribute. Making these different perspectives play well together, whether in academic research or commercial firms, is essential to adding to our knowledge through the

(continued)

application of data science and is often negotiated in practice by the careful development and management of trust (Passi and Jackson 2018). Misplaced or unearned authority, or a data science that is uninterested in the knowledge and practices of real-world actors ("just give me the data"), is the enemy of this process. So are instances in which data science is brought in (for example, by management) to overrule local knowledge. One example is the essential knowledge and experience of experts in the domain. To paraphrase a point sometimes attributed to Winston Churchill, what is needed is a data science "on tap, not on top."

Ordinary, humble, fallible, and collaborative: this would be a human-centered data science worthy of the name.

A completely different approach to human-centered data science has emerged from the social science fields, such as science and technology studies, involving studying the people who produce data science. What are the sociological factors at play in data science teams? What constitutes a successful data science collaboration? Reflecting on these types of questions can lead to changes in the human process of data science work and ultimately to more fruitful collaborations and better results.

Another way to do human-centered data science is to combine approaches. Small-scale qualitative approaches to data collection and analysis offer researchers the opportunity to obtain very rich, deep insights about specific phenomena—often in a very bounded or limited context (Zheng et al. 2015). Such studies often face challenges related to generalization, extension, verification, and validation. They also face problems of scale. It is possible to interview 100 people but very hard to interview one million. On the other hand, large-scale quantitative approaches to data collection and analysis give the opportunity to look at broad datasets, but the insights gleaned are often much shallower, lacking the rich detail associated with deep study (Green, Arias-Hernandez, and Fisher 2014). "Big data needs thick data," as one anthropologist termed it (Wang 2016). There are now many people, including the authors of this book, who advocate for combining the power of data science tools to understand humans at scale with ways of understanding human behavior at depth through qualitative approaches that can provide powerful insights (Muller et al. 2016; Baumer et al. 2017).

Human-centered data science draws on work from both qualitative and quantitative traditions, involving practitioners with training in computer science, statistics, or social science. Examples include work that has integrated quantitative research methods into qualitative research workflows (Brooks et al. 2013; Goel and Helms 2014; Xing et al. 2015). Online, digital, or virtual ethnography has gained widespread adoption as qualitative researchers adapt traditional ethnographic methods to online spaces (Daniels, Gregory, and Cottom 2016; Markham and Baym 2009; Murthy 2011). Computational social scientists—that is, researchers primarily in social science areas who develop and use computational methodologies to ask and answer social science questions—have found significant recent success in developing computational methodologies for large-scale social data that account for a degree of contextual reasoning within analysis.

Human-centered data science also encompasses the process of creating data science tools that integrate seamlessly into the sociotechnical ecosystem of the domain they are designed for. Such tools have often demonstrated the greatest success. One well-known example is iPython (later Jupyter), first developed by Fernando Pérez in a human-centered fashion specifically to ease scientists' workload (Pérez and Granger 2007). Human-centered design is particularly effective in the development of software for analyzing large datasets (Aragon and Poon 2007; Aragon, Poon, and Silva 2009; Faiola and Newlon 2011; Poon et al. 2008).

Among the many unanswered questions surrounding human-centered data science are issues of sampling, selection, and privacy. What are the ethical questions raised by processing vast datasets? How should we treat the workers who do necessary tasks on crowd-work platforms? Who owns personal medical data—the company whose machines and software collect it, the medical practitioner who interprets it, or the patient who generates it? Can design or other skills often considered to be the province of an individual human be effectively crowdsourced (Bean and Rosner 2014; Lasecki et al. 2015)? What policies do we need to develop to protect human rights in this new age of "big data" (Gray and Suri 2019)? Questions such as these are legion, and we are only beginning to explore the territory of potential answers.

About This Book: Themes

First, we would like to draw your attention to five recurring themes that are developed throughout the book. We ask you to reflect on each of these themes and consider how they are used as you read.

Human-Centered Data Science as Ethical Responsibility: "The Data Made Me Do It"

One of the difficulties in dealing with the "data deluge" is a facile assumption that the data can tell us everything—that it is unbiased, neutral, and somehow possessing wisdom far beyond the human. "If it's big enough, it contains everything."

Our approach puts human responsibility at the center of data science. People are involved at every stage of the cycle of collecting, cleaning, analyzing, and communicating data science results. Each stage presents a series of choices, and these choices matter for the responsible and ethical use of data.

Human-Centered Data Science as Looking in the Right Places: The Streetlight Effect

The streetlight effect is a kind of observational bias where people only search for something where it is easy to look. It refers to a joke that apparently dates to the 1920s. A police officer sees a person on hands and knees searching the ground around a streetlight at midnight and asks what they're doing.

"I'm looking for my keys."

The officer helps for a few minutes, doesn't find anything, and eventually asks the person if they're sure the keys were lost near the streetlight.

"No, I lost them across the street somewhere."

"Then why look here?" asks the irritated officer.

"The light is much better here."

The streetlight effect explains, perhaps, why many researchers (including, we have to admit, some of us) have turned to Twitter to study social phenomena. People take advantage of datasets that are easy to access or easy to convert into simple data structures for analysis. Research shows that Twitter data has serious limitations as a representation of public opinion. But because it is public, easily available, and vast in quantity, hundreds of research papers have been published using Twitter data. Certainly, data science can do better than look under the streetlight. A human-centered approach to data science asks where can we look first, before looking under the light of the easily available dataset.

Human-Centered Data Science as Collective Practice: We Are All Problem Seekers

Our approach to data science holds that many people can be empowered with data skills. You are reading this book and that is a start. We work with a range of communities, including self-trackers, child welfare agencies, community crisis response activists, astrophysicists, architects, journalists, nurses, pharmacists, and citizen or community scientists; clearly, data science in the human-centered approach is not a toolset reserved for the elite and the powerful. Our research shows us that working with people who have deep inside knowledge of the problems we are trying to solve helps improve our practice as data scientists. A human-centered approach figures out what needs to be known from the situation to create better models, more responsible data science pipelines, and more capacity for using data science tools—responsibly and ethically—to benefit people. We are encouraged by the Community Data Science Workshops and Urban Data Science, which host free and open meetings to train others (Hill et al. 2017; Rokem et al. 2015). We are inspired by people in the Data Science for Social Good (DSSG) movement who come together to learn experimentation and data science techniques from one another.

Human-Centered Data Science as Communication: We Are Communicators and Storytellers

Data science tools and methods are complex and multifaceted. We think of them as analytic lenses through which we look at the world or craft our version of the world. We also think of them as tools for reflection—on the data, on the tools, and on our own evolving understanding of our own ways of thinking about data. We use data science to tell stories about data and people who are affected by our choices of methods to analyze that data.

Human-Centered Data Science as Action: Make a World Where We Want to Live

Some of the authors of this book are software developers and data scientists with practical experience in industry. We look at the consequences of our technology, and we want to build technologies that create a world that we want to live in. For example, we would not want to build facial recognition technology that leads to false positives and wrongful convictions. We would not want to live in a world where digital surveillance is an everyday presence. As technology workers, we want to work toward a better future. Within human-computer interaction, this approach is sometimes referred to as *prefigurative* design and action, emphasizing both design practices and design outcomes that correspond to the future that we collectively envision (Asad 2019; Strohmayer, Clamen, and Laing 2019; Williams and Boyd 2019).

About This Book: Stories, Audience, and Our Purpose

We now shift gears from describing human-centered data science to helping you make the best use of this book.

Stories and Case Studies

We believe in the power of stories. In each chapter we use stories about data science practice to illustrate the main themes. Throughout the book, we present short case studies to illustrate some of the ramifications of human-centered data science. These case studies bring multiple authors into this book to present real-world examples of how to use human-centered data science, critique data science, and work with multiple communities.

Because we want this book to help people have a real-world impact, at the end of every chapter, we provide a set of recommendations and things to consider while doing a data science project. We also list recommended readings that go into more depth on the topics covered in each chapter.

Who This Book Is For

This book is addressed to people doing data science, learning data science, or managing data scientists. We imagine you, the reader, to be someone hoping to learn more about data science—either in a formal course or on your own, either as a student or a practitioner. We provide a brief overview and easily understandable explanation of many of the common statistical and algorithmic data science techniques to emphasize how a human-centered approach can enhance each one. You do not need any specialized knowledge in data science, computer science, or social science to learn from and benefit from this book, although we summarize and discuss many decades of research and experience from each of those fields. We don't intend this book to teach you how to do the latest techniques in data science. However, we think, modestly, that you can't do good data science without the practices that we cover here.

Why We Wrote This Book

Universities, businesses, and governments have rushed to train millions of people in the computational and statistical techniques necessary to process and extract insights from the vast amounts of structured and unstructured data. This computational turn toward so-called big data means the proliferation of more types of data generated and collected from a variety of sources. However, in the process, the social context and ethical considerations of data collection, analysis, use, and dissemination have often been overlooked. Many well-documented cases show how some approaches to data science can lead to severe ethical transgressions and significant harm, social bias, and inequality. And yet, from the purely computational perspective, many of these issues and complications may be hard to foresee, especially for aspiring data scientists who have no background in the ethics of data science from a human-centered perspective.

We have high hopes that this book can become a practical manual for data science practitioners who want to change the world. We do not say this lightly. We believe in the power of data science to help people discover new things, solve urgent challenges, create new services, and make things more efficient and better. We believe that people working in data science have a responsibility to the people who are affected by the results of their data, to the people whose data they use, to the people they work with and for, and to communities that may make use of their results. We believe that people want to—and can— work ethically and responsibly. Our approach to data science understands that people are often the source of data; that people do the work that it takes to create, label, source, and analyze data; that people are the audience for our results and rely on our ability to clearly communicate what we discover in our projects. We believe that data science can be done better if it is done in a way that is both technically rigorous and addresses concerns about bias, ethics, and inclusion. In that sense, we—a group of researchers and data scientists in universities and industry—are on a mission to help bring that urgency of human-centeredness to the next generation of leaders and practitioners in data science.

In this book, we explore how data-driven and qualitative research can be integrated to address complex questions in diverse areas, including but not limited to social computing; urban, health, or crisis informatics; and scientific, business, policy, technical, and other fields.

By training people in the human-centered data science methods described in this book, we hope to address concerns about the social impacts of large-scale data. This book can be used alongside a textbook for students of data science, both undergraduates and graduates. It can also be a useful handbook for professionals seeking to learn more about the responsible, effective, and human-centered use of data science to process the ever-growing quantity of human-generated data in the world today.

Book Outline

In chapter 2, we introduce the data science cycle, or the stages involved in a typical data science project. We address data collection, cleaning, feature engineering, analysis, representation, bias detection/mitigation, and distribution, with attention to the ethical and human-centered considerations that inform each step. We emphasize a cyclical, rather than linear, trajectory for data. We then describe several techniques commonly employed in data science and discuss the most popular tools used to carry out each technique. Our goal is not to teach these topics in-depth but to show how human-centered approaches might be useful for improving them.

Chapter 3 foregrounds the social aspects of human-centered data science work, exploring the relationships among various stakeholders in data science work: subjects, researchers, and their audiences. We show how people who work in data science usually have to intervene between "the data" and "the model" in data science pipelines. We discuss how data can represent both voluntary participants and unknowing subjects, and we address some of the ethical issues involved in data collection. We also examine populations that have historically been most vulnerable to exploitation by data science projects, and we think about how data scientists might interrogate both their methods and their data with regard to protection of privacy.

In chapter 4, we present a high-level overview of commonly used tools in the data science toolkit: machine learning, statistical analysis, automated tools for constructing data science solutions, visualization tools, and others, retaining a human-centered focus.

These tools and techniques form the core of textbooks that our book aims to supplement. We spend no more than a few paragraphs on each of these techniques, with a focus on advice about when each might be most appropriate to a given research project. The concepts in chapter 4 may help anchor some readers who are new to data science in the tools and techniques of data science. Reviewing these concepts also gives us the opportunity to reflect on the human-centered challenges of these tools.

Chapter 5 focuses specifically on human-centered approaches to asking and answering data science problems. We start with the history of ethical research design to show how human-centered data science differs from other data science practice. Human-centered data science starts by formulating a meaningful question, considers issues of ethics and fairness, designs projects that others can easily build on, and incorporates reflexivity into the process. Many of the suggestions we make in chapter 5 urge you to think about your project's impact on people represented in your dataset, people whose lives may be affected by the results of your analysis, people who might want to reuse your pipeline or just view your results, people who may reuse your results, and even yourself at a later point in time.

In chapter 6, we extend the human-centered data science toolkit by examining how methods from other fields might be integrated and combined in innovative ways with data science. We draw on the rich tradition of more than a century of social science research into human behavior to explore recent research on new ways of merging computational tools with the methods that people have developed to interpret and transform our social world.

In chapter 7, we address the multiple types of collaborations that are needed for human-centered data science. Most data science projects are conducted in teams, and we consider how collaboration, especially with people from other disciplines, is an integral part of the process. We also discuss the different roles people take on in data science teams throughout the project lifetime. We examine various ways data scientists can collaborate with AI in their projects. While considering how collaboration works in organizations, we delve more deeply into the notion of a stakeholder, giving you tools for untangling complex webs of interaction and power in large organizations. Finally, this chapter emphasizes the importance and often complicated nature of working with the communities whose data you plan to use or whose problems you intend to solve with data science projects.

In chapter 8, we consider data storytelling and offer suggestions for how to translate human-centered data science principles into action. We draw on the rich literature of storytelling with special focus on why stories matter for data science practice. We then focus on how visualizations work as a great storytelling medium and provide recommendations for how data scientists can use visualizations (and storytelling, more broadly) to connect with various communities, including business leaders, experts, academics, researchers, and policymakers.

Chapter 9 addresses the future of human-centered data science as a discipline and discusses how researchers and practitioners can be advocates for human-centered data science practices among students, business leaders, policymakers, researchers in other domains, and the public. We review the five cross-cutting themes that we first introduced in chapter 1 and reflect on the examples and challenges of each. We conclude with a call to action posed by the poet Adrienne Rich (1986)—"With whom do you believe your lot is cast?"— and look ahead to how we all can shape the future of data science.

Who We Are

The authors are a diverse group of researchers with long-term experience in human-centered data science. In February 2016, four of us came together at a workshop titled "Developing a Research Agenda for Human-Centered Data Science" at the Computer-Supported Cooperative Work and Social Computing (CSCW 2016) conference. Although we had all been working in this area for a few years, this workshop served as a catalyst for us to develop focused research to build the field.

Cecilia Aragon, originally trained as a mathematician and computer scientist, has been conducting qualitative and quantitative research in human-centered data science for over a decade in academia as a professor at the University of Washington (UW), after fifteen years of hands-on experience as a data scientist and software developer in industry. She coined the term "human-centered data science" and organized the first workshop on the topic at the 2016 CSCW conference. She is Professor and Director of the Human-Centered Data Science Lab at UW and a strong advocate for the use of human-centered techniques throughout data science. As founding faculty director of the interdisciplinary data science master's program at the University of Washington, she developed the original curriculum for its course in human-centered data science.

Shion Guha has formal academic training in economics, statistics, and information science and is a professor of human-centered data science in the Faculty of Information at the University of Toronto. He uses computational and qualitative methods to examine how data-driven algorithms are designed, deployed, and evaluated in public services, particularly in the child welfare and criminal justice systems. He is building the undergraduate and graduate programs as well as curriculum in human-centered data science where questions of ethics, inequalities, and social justice take precedence in academic discussions.

Marina Kogan is a professor in the School of Computing at the University of Utah. Her research focuses on how people self-organize and problem-solve on social media during disasters. Her methodological focus is on developing methods that attempt to both harness the power of computational techniques and account for the highly contextual nature of the social activity in crisis. She extends and develops human-centered versions of network science models, natural language processing (NLP) techniques, and probabilistic models.

Michael Muller is enthusiastic about working with users (including data scientists and *their* users) for increased mutual understanding and collective action to make better outcomes for everyone. He has researched data science work practices at IBM Research AI. His research methods span qualitative and quantitative approaches, including the grounded theory analysis in his 2019 paper on data science workers and quantitative survey analysis in his 2020 paper on collaboration patterns in data science teams. Michael's background includes extensive and sometimes passionate work in participatory design, organizational social media, and allyship for social justice.

Gina Neff leads qualitative research teams on data science studies, looking at how data science is made in practice in industry settings. She directs the Minderoo Centre for Technology and Democracy at Cambridge. Her research focuses on work and collaboration, and she uses these insights to advise universities, startups, and nonprofit research organizations, including Data and Society and AI Now.

2

The Data Science Cycle

The invitation was simple enough: meet for a coffee in the coolest café in a neighborhood known for its density of tech company headquarters. One of the authors (Gina) had met someone at a health innovation conference who had experience in luxury consumer goods and left to work on a digital watch that would serve to gather data about users' daily exercise and movements. The question he asked of her was simple and yet hard to answer, "What are we going to do with all this data?"

—Neff and Nafus 2016

It is the way data science often works: start with the dataset, then figure out something to do with it. Social scientists start with the question, then figure out the data needed to answer it. From wireless sensors to mobile phone geolocation to social media, society is awash in "all this data."

We anticipate that people learning data science are asking the same question: What are we going to do with all this data? This question has motivated an explosion of data science opportunities and jobs. And it is one of the ways the data science cycle begins in many industries. Our goal in this chapter is to introduce this standard cycle, which is covered in-depth in other textbooks. We present it here to show readers the typical structure for the data science process and at the same time use it as a springboard to discuss the human-centered approaches that we focus on in this book. In chapter 3 we will examine many assumptions inherent in the traditional data science cycle. This chapter lays the foundation by walking you through the standard cycle as is currently practiced by many data scientists.

The data science cycle is often imagined as a series of sequential, interconnected steps. It is a cycle because there is an element of self-evaluation and feedback in every step that circles back to the initial stage of asking questions (see figure 2.1).

A typical process starts with question or problem formulation, then goes through data collection, wrangling, cleaning, modeling, and finally representation, evaluation, and interpretation of results. Realistically, self-evaluation and feedback may exist at each step. We expand on each of these elements in the sections that follow, then conclude the chapter by distinguishing between models and pipelines and how they fit into a data science cycle.

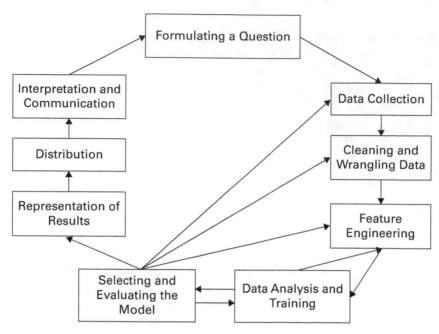

Figure 2.1
The data science cycle.

Elements of the Data Science Cycle

One "standard" data science cycle lists nine elements (figure 2.1). The first step is often defined as *formulating a question*, although this step may not come first or the question may be revised multiple times. *Collecting data* can be accomplished in many ways. *Data wrangling and cleaning* refers to the unexpectedly lengthy and difficult process of fina-gling the data into a state that makes it easier to process, analyze, and visualize. *Feature engineering* involves selecting and extracting the attributes or features of the data that will be included in the model or algorithm, and sometimes making new combinations of features (e.g., ratios) or reassigning the labels of features ("classes") based on data outside the dataset. Next comes *data analysis and training* of the model, often on a subset of the full dataset. The first time through this step you will focus on data analysis for later train-ing. Selecting labels or "ground truth values" for this training comes with its own set of decisions and potential pitfalls. Next, the iterative nature of the cycle becomes clearer as you *select and evaluate the model* and perhaps iterate on model selection. You will likely also return to the previous step and continue training the model. After that comes the decision of how to *represent the results of the data analysis* (i.e., communicate the results or provide for further exploration) to your audience—customers, decision makers, or any-one affected by the outcome of the analysis. *Distributing* the results of your analysis to others through publication is an important component of the cycle. Finally, *interpretation and communication* of your results to other people, through discussion or presentation, is a fundamental data science task. Throughout this cycle, the *iteration and feedback* pro-cess is critical. Iteration is not a setback but rather a process that deepens and strengthens the quality of your end results.

Formulating a Question

Most versions of the data science cycle put formulating the question or stating the problem as the first step in the process. We agree and suggest that data scientists consider asking a specific question they would like to answer through the data science process, as opposed to only starting with a specific dataset. Starting with a dataset may limit the types of questions we are able to ask and answer, and it curtails imagination for what other types of data might be appropriate and may be even better suited to answering a particular question. Starting with the data puts us in the position of not proactively looking for an answer to a question or a solution to a problem; instead, we are reacting to the specifics of a particular dataset. Depending on how the dataset was collected, how the variables in it were measured, and by whom, the results of your analysis may be strongly skewed or even distorted. In addition, people often take the path of least resistance, choosing to work with datasets that are easy to get or provide easily quantifiable data. But these may not in fact be the best datasets for answering a particular question—just the most convenient ones. We want our dataset to be in service of our question, not the other way around.

- How have you approached your data science projects? Did you "start with the data" and try to "find a good question" from the data? Or did you "start with the question" and try to find the right data to answer that question? What strengths and weaknesses have you encountered with each of these approaches? What other approaches have you experienced?

Another concern with starting with the dataset is that it is harder to know whether you are really measuring what you intend to measure. Some disciplines (e.g., psychology) call this a *measurement plan*—laying out the steps to quantify and assess the concepts pertaining to your question. For example, in health-related machine learning we may want to predict mental health indicators to be able to help people in real time. But how do we know if specific behaviors we are measuring are indeed good indicators or proxies for, say, depression? Other disciplines call this *internal validity*: How valid is our measurement for capturing some concept? Starting with formulating a question will also allow you to think deeply about how you measure different concepts within it, instead of choosing variables that might be poor proxies but are part of an easily accessible dataset.

Collecting Data

The next step is usually data collection. There are many ways of collecting data: downloading existing datasets that have been curated by others (individuals and organizations), collecting data from surveys, capturing activity on various software systems through their logs (such as credit card transactions or mobile phone records), scraping data from the web, using application programming interfaces (APIs) to download structured data directly from websites or apps, collecting scientific data from sensors, and many others.

If you choose to work with an existing dataset that has been curated by others, first you need to learn how the data was collected and what additional information you may need to be able to make use of the data: what variables are in the dataset (usually as columns in a table), what does each variable measure and what does the metadata mean, who measured these things and how, and how reliable and trustworthy are these sources of information (the people who compiled the data). You also might need to request the data from the people who collected it. In this case, you would need to negotiate access.

If you scrape the data from the web, you need to consider the terms of service of the site you are interested in and how appropriate it is to harvest the data. In that case, you will want to consider the potential harm of disrupting the site with repeated requests, the dangers of combining the resulting data with other publicly available datasets that might lead to deanonymization, and other possibilities (Fiesler, Lampe, and Bruckman 2016). For a deeper discussion of legality and ethics of web scraping, consider reading Krotov and Silva (2018). You should be aware that there are diverse views on ethics versus legality of working with website data (Bruckman et al. 2017), and that there are different expectations and legal frameworks in different regions of the world (Voigt and Von dem Bussche 2017).

As you might have noticed, each type of data collection entails questions of values and ethics: what data is needed and from where should it be collected, whose worldviews are reflected in particular datasets, how are various concepts being measured and by whom.

- Laws and expectations for data access vary by country and culture. What have you had to do to make sure that your data access was ethical and legal in your location and institution? Did you experience the rules as helpful, obstructive, or protective? If they were protective, whom were they protecting?

This relates to the problem of *justifying our data collection*. In data science, especially if we work in industry, we may have access to many different types of data that could potentially help us answer a question. Imagine a data scientist at Facebook aiming to answer a question about people's preferences for certain types of Facebook Groups (let's say around sports). It might be tempting for them to gather all the possible types of data that Facebook has on people who participate in these types of groups: data from groups themselves, their conversations and reactions within the groups, but also their conversations and reactions outside the groups, their private messages, and so on. There are a lot of data sources to consider. To justify the data collection, the data scientists (individually or within teams) should deeply interrogate what types of data they actually need for this analysis, as opposed to using everything that is available just because it is there. This interrogation—acting as your own adversary—will lead to more purposeful and intentional use of data.

In industry applications, when data scientists start with a particular question of interest to their clients, they may have to switch the dataset they rely on multiple times, based on the availability of the data, its granularity, and ease of access. This sometimes means that they repeatedly trade more precise and granular sources they had in mind for less accurate approximation of the behavior in the data that is more realistic to obtain. This means that the *measurement plan* they had at the beginning of their analysis also keeps changing. In this case, the data scientists need to be even more vigilant about to what degree they can answer their original question and with what level of confidence. These limitations need to be explicitly communicated along with the results of the analysis.

For example, in a study of crowding in a very large urban transit system, researchers first used surveys but needed to obtain data around the clock and in more transit stations

(Muller, Lange et al. 2019). They tried to use the closed-circuit TV cameras in the transit stations to estimate crowd size, but discovered that not all transit stations had cameras. They finally engaged in complex negotiations with mobile phone carriers to use smartphone signals as their estimate of crowd size and crowding. At each step in this process, they had to make new compromises regarding the cost of obtaining data and their ability to measure what they needed to know from the new data sources.

- Have you ever had to switch datasets during a project? How did you decide that this was necessary? What was the "benefit" of switching? What was the "cost" of switching? How much of your work did you have to redo for the new dataset? Did the new dataset produce better or more complete analyses?

As we mentioned with web scraping, combining datasets can be a powerful tool, but it can also pose significant dangers. One such issue is the potential for triangulation (use of data from multiple datasets), which can lead to deanonymization (privacy violations) when combining multiple datasets. In addition, even individual datasets can cause significant breaches of privacy for people who are represented in them. For example, some Twitter users geolocate their tweets with precise coordinates. Aggregating those tweets over time makes it rather easy to estimate their home locations and other geographic data, with strong implications for privacy (Valentino-DeVries et al. 2018). Chapter 3 discusses in more detail the various privacy issues that can arise from individual and multiple datasets.

Cleaning and Wrangling Data

Data wrangling refers to the often-messy work of getting data ready for analysis. Wrangling can take as much as 80 to 90 percent of the time and effort of a data science project (Rattenbury et al. 2017). It involves cleaning the data—for example, deciding how to deal with the missing values (i.e., the data cells that are either empty or automatically generated in place of empty cells because the actual observations are missing). Missing values can be handled in a variety of ways, including deleting certain data points, deleting entire variables (columns in a spreadsheet or dataframe), or imputing the values—replacing the missing values with different types of synthetic (computed) data. These missing-value imputation methods involve human decisions about what is appropriate for different types of variables and datasets. We must be especially careful here because either removing or imputing missing values can change the distributions of the imputed or other variables in the dataset, since the variables are often not independent. For example, dropping people who did not finish the survey (missing data) may affect the gender distribution in the dataset if the missing data was because, say, the survey included a question that women tended to find difficult or offensive to answer, resulting in fewer women completing it.

Another aspect of cleaning the data is dealing with *outliers*. These are data points that differ significantly from the other points, which might indicate an experimental error or a data collection error. Hence, outliers are often removed from the data. However, choices of how to handle outliers can affect who or what gets left out of the dataset. For example, if we apply the norms of European-descended people to a column about body height, then we may exclude as "outliers" certain Indigenous populations from Africa. These data points may in fact be important and demonstrate the variability in the data and removing them would make the dataset less representative. Thus, decisions about missing values and outliers have to be considered carefully.

Real-world data also may be structured in a way that is not conducive to analysis. That is usually called messy data (Kery 2018), and we can apply certain principles to transform it into "tidy data" (Wickham and Grolemund 2016). The three main principles of tidy data are that each variable forms a column, each observation forms a row, and each type of observational unit forms a table (Wickham 2014). This type of structure in the data makes it more interpretable and easier to work with. However, real datasets often violate these principles in some familiar ways:

- Column headers are values, not variable names.
- Multiple variables are stored in one column.
- Variables are stored in both rows and columns.
- Multiple types of observational units are stored in the same table.
- A single observational unit is stored in multiple tables. (Wickham 2014)

This is another stage where we need to interrogate our datasets to make sure they are ready for analysis, but we also want to make sure that the transformations we apply to them do not exclude certain types of data points from analysis just because they are more difficult to convert into a tidy format. For example, if we use a simple binary female/male definition of "gender," then we may inadvertently exclude people whose gender identities do not conform to that binary.

We want to emphasize that spending the time to understand, clean, and tidy the data is very much part of data science work, even though it might sometimes not feel that way. Reconstructing the documentation and clarifying the metadata are important in knowing and understanding your data, so you can correctly perform the next steps in the cycle. Wrangling is a well-known problem, and you can find detailed methods, practices, and tools to help with this work (Kery et al. 2018; Rattenbury et al. 2017).

- Many data science workers have "wrangling stories" of particularly difficult or time-consuming cases. Did you experience a challenge in data wrangling? What made it difficult? In retrospect, what could you and your team have done to make it easier? What compromises did you have to make? With the benefit of hindsight, what would you have done differently?

Feature Engineering

Feature engineering is the practice of selecting existing variables (or features) in the dataset for inclusion in the model, as well as producing new combined features that capture additional aspects of the dataset. Such new features may be desirable as some combinations of existing variables may provide more predictive power for modeling. Feature engineering is also often performed for *dimensionality reduction*: reducing the number of variables included in the model because we often cannot afford computationally to include all the columns we have. Another reason to perform feature engineering is to split multi-class variables (categorical variables with more than two classes) into several simpler categorical variables that only take on the value of zero or one, called *dummy variables*. There are also automatic ways of splitting multiclass variables into a weighted version of dummy variables; examples are effect coding in statistics and one-hot encoding in machine learning (Kugler, Dziak, and Trail 2018).

While features are intended to be a representation of things in the real world, some things are easier to represent as variables than others. Easily measured and quantified things are readily turned into variables or features. Human relationships, traditions, and local ways of making sense of the world may be harder to represent as quantifiable variables. And even when we work hard to find a way to measure these things and make them "legible," the resulting variables often reflect only a narrow view, a single aspect of these complex and multifaceted human endeavors (Scott 2020). For example, how do we represent how friendly and cordial a neighborhood is? It is not a well-structured, easily measurable aspect of a neighborhood, unlike the number of houses or streets. We may try to measure friendliness by the number of neighbors who say hello to each other on the street or welcome new neighbors with baked goods, or the number of children who have friends on their street. Clearly, all these ways of turning neighborhood friendliness into a quantifiable variable are only partial, one-sided representations of the complexity of neighborhood life, and we have to think carefully about which of these (or what combinations of them) are a better fit for our investigation.

Finally, we sometimes create features as a way of "controlling" or "normalizing" one variable with another. For example, if we want to include a predictive feature about how "characteristic" a certain word is in a document, then we might want to count the number of instances of that word (using natural language processing) and then divide by the number of words in the document to compute a percentage. Suppose our target word occurs ten times in a document. If the document is fifty words long, then the target word is probably *very* characteristic of that document. If the document is 10,000 words long, then the count of ten instances makes the target word less characteristic of the longer document. One of the tricky aspects of data science is deciding what is a "fair" way to normalize important numbers.

Since the new features are generated through various combinations and transformations of existing variables, feature engineering is an inherently human decision, even though there are often rules of thumb on what new combinations may be useful for different application domains (see chapter 3 for a discussion of feature engineering as the design of data). In addition, feature engineering generates and selects the features we consider using in the model. Here again, we need to think critically about these new features—whether they measure what we intend to measure, how they relate to our measurement plan, and whether they introduce any unintended bias. For example, using a geometric mean of three existing features might seem promising for some computational reasons, but what would it mean in terms of the concepts of interest, the question we are trying to answer? Would the simple arithmetic mean (average) be a better measure? Or the median?

- Have you had to construct new features from your data? Did you do this alone or with others? How did you decide which new features to engineer? How did you test them to see if they were useful and informative? Did you evaluate the possibility that the new features might have introduced bias?

Data Analysis and Training

In some data science techniques, a *training dataset* is used to help create and fine-tune the model. This requires *labeled* data: a subset of data points annotated with labels with categories or classes of interest. Labeling may also be called *annotating*. The labels or annotations are used as the *ground truth*—an accurate representation of the world from which the model is supposed to learn to generate labels when presented with new, unlabeled data (see case study 2.1 for the importance and implications of labeling the data). The choices made in how data science models are trained have implications for the results. How the training dataset is selected is an important choice. The size of the training data influences the results, but so does its representativeness.

We need to carefully consider who and what are included in the training set, as that will influence what the model finds in the data. Thus, types of social values and choices that are included in the training data will have implications for the model. A model trained to identify shoes but trained only on athletic shoes will miss other types of shoes. Similarly, a training dataset that relies on cases drawn from only one neighborhood may miss key factors from the rest of the city. The main consideration here is that training data should represent the phenomenon across all relevant diversity measures. This can become a difficult decision. If you are sampling across people, is race an important type of diversity for the phenomenon that you are working on? Is gender identity? What about age? You may need to look at other, similar projects to see which diversity variables they have used. You will also need to exercise your own good judgment and be attentive to your own biases, and you may want to consult with a diverse group of other people who have worked on these questions.

In addition, assigning labels can often be a difficult and costly process, and there are many decisions about strategies for data labeling (who labels it and how). These issues are discussed in more detail in chapter 3. Once we have a labeled set of training data, we train the model based on these labels. In practice, we often train multiple models to determine which type of model represents the data best, including a set of model parameters. One important concern is to avoid *overfitting*, or creating a model that fits the existing data so closely that it is not generalizable. How to do this is discussed in chapter 4.

Selecting and Evaluating the Model

There are different ways to select the best model for the job. In data science, we often use measures of accuracy. When making a prediction, the simplest measure we often use is called *classification accuracy*. It measures the proportion of the correct predictions out of all the predictions we made. It works well if there are equal numbers of cases in each class but can be misleading if classes are mismatched in size. For example, if 95 percent of our data had green labels and 5 percent had red, the model simply predicting all the labels to be green would be 95 percent accurate. But of course, such a model is not very useful, as it would not perform well on other datasets.

Overall accuracy is also not very useful, as we want to ensure that the model does well for every class. For example, if the model predicts men's behavior with very high accuracy but predicts the behavior of women somewhat poorly, the overall accuracy would still be rather high. Instead, we are looking for a model that has high accuracy for each class.

Case Study 2.1
Combining Knowledge in Data Science about Mental Health
Stevie Chancellor, Northwestern University

I build machine learning and data science tools to identify high-risk mental health behaviors in online communities. By high-risk, I mean behaviors like self-injury, suicidal ideation, and disordered eating and exercise behaviors. My aim is to build models on millions of public posts that can support better decisions about this data—like when to make an intervention. Online communities are an important source of support for those with mental illness, and I hope to support these people and those who care for them.

An important piece of my human-centered data science work is *construct validity*: that is, getting high-quality, accurate labels for my data. For me, a label might be a diagnosis of depression or quantified risk evaluation. Because I use labeled data to build reliable and precise models, it is essential that the labels are as close as possible to the corresponding clinical concept. For example, when we study depression, we should look at diagnostic criteria and clinical definitions. This is critical to the success of my work: if my research team's data labels are not valid, we could misclassify someone as suffering from a disorder, or we risk missing someone who needs help.

To support strong construct validity, I work with subject experts who are familiar with the clinical concepts my labels are trying to replicate. That often means working with doctors and medical researchers—but sometimes it also means moderators or community members themselves. I talk to many different "experts" with different experiences to help triangulate these concepts—triangulation is a research practice that uses information from multiple sources to converge on a conclusion. This makes my research very collaborative and fun and helps ensure the labeled data is a valid representation of mental illness.

In one project, we explored high-risk mental illness behaviors on Instagram. Severe mental illness is defined as major cognitive, judgment, and behavior impairments that make it hard to take part in day-to-day life. In our study, we focused on suicidal ideation, extreme weight control behaviors (like binging and purging, overexercising, and extreme food restrictions), and self-injury.

But we are computer scientists, not doctors—how are we going to understand this phenomenon, stay true to its clinical definitions, and ensure construct validity? To accomplish this, we worked with two medical researchers to identify signs of mental illness severity in eating disorder communities. Stephanie Zerwas and Erica Goodman are two clinical researchers who actively see patients as well as conduct research on eating disorders online. They were true research partners on this project and led on key aspects of our approach. For instance, they evolved our labeling guidelines from simply recording whether the post in the sample was "severe" to a more nuanced three-tiered approach that reflected their clinical reasoning. We then developed an annotation system using both clinical judgments and computational linguistics that annotated 26 million posts on Instagram for mental illness severity.

Working with domain experts like Stephanie and Erica is a crucial part of my research for a few reasons. First, these researchers make my models more robust because the experts think through examples I would never know as a computer science person. Accurately labeled rare and surprising cases are key to developing machine learning models with high sensitivity. Working with domain experts gives me confidence that the data labels we have are closer to a real idea of mental illness severity that may be used in a clinic or by a doctor, since Stephanie and Erica see patients in addition to conducting research. The second benefit is true of all collaborative research: I enjoy working with smart people, and their approaches encourage me to think about my problems in new ways. Our work together helps them consider theirs in novel ways, too.

(continued)

Case Study 2.1 (continued)

> I've used this approach to great success in studies about other behaviors like eating disorders and opioid addiction. Recently, I have been working with domain experts about suicidal ideation in online communities with the US Centers for Disease Control and Prevention to help them develop monitoring for suicide risk. I aim to involve experts in my work from the very beginning of my projects and view this as a core component of human-centered data science. "Construct validity" was the term I used at the outset, but ultimately this means not losing the humans represented by labeled training data. Keeping humans, and the experts that know about the human phenomena we're studying, at the center of our work helps ensure that the results we generate are valid, useful, and grounded in appropriate domain expertise for the people we are trying to help.

In addition, we would like to know exactly in what way the model fails in its prediction and where it does well. A *confusion matrix* presents the full performance of the model in a matrix form. Assuming a binary classification problem, we have two real classes in the data: 0 and 1. It is common to use 1 for the presence of something (or a positive outcome) and 0 for its absence. We also have our classification predictions, which may or may not match the true class of each sample.

The confusion matrix (figure 2.2) cross-tabulates the actual class labels (rows) and the labels predicted by our model (columns). The resulting cells contain frequencies of four important measures. Suppose that we are trying to predict outcomes that can be represented numerically as 0 or 1, where 0 represents a negative outcome and 1 positive. Then we can define the following measures:

- True positives (TP): the cases in which we predicted 1 and the actual output was also 1.
- True negatives (TN): the cases in which we predicted 0 and the actual output was 0.
- False positives (FP): the cases in which we predicted 1 and the actual output was 0.
- False negatives (FN): the cases in which we predicted 0 and the actual output was 1.

False positives and negatives are important measures for diagnosing the weaknesses of our model and determining where it failed and how.

Another way to evaluate the success of a model is through the metrics of precision and recall. *Precision* is the number of correct positive results divided by the number of positive results predicted by the classifier. This is the portion of true positives out of all the positive results predicted by the model (true and false). It signifies how precise the model is—that is, how many positive instances it classifies correctly. *Recall* is the number of correct positive results divided by the total number of relevant samples. Essentially, this is the proportion of true positives out of all the predictions. It signifies how robust the model is (i.e., it doesn't miss a significant number of positive instances). High precision but lower recall means an accurate model, which at the same time misses a large number of instances that are difficult to classify.

One of Michael's participants described a project to determine whether the audiogram of a human cough was a symptom of a viral infection or a bacterial infection (Muller,

Actual Values

	Positive (1)	Negative (0)
Positive (1)	True Positive	False Positive
Negative (0)	False Negative	True Negative

Predicted Values

Figure 2.2
Confusion matrix.

Lange et al. 2019). Coughs were being measured in low-resourced field clinics, with very limited supplies. The goal was to make an audio recording of a cough using a smartphone, then use it to make a diagnosis. Precision tells us about the "positive" result—in which case the cough is due to bacterial infection and we should treat it with an antibiotic. Of course, we want high precision in order to save lives. Recall tells us about the "negative" result in which the cough is due to a virus; in this case, an antibiotic would not be an effective treatment. Because the field clinics had a limited supply of antibiotics, it was also important not to deplete that limited supply when the antibiotic would not help the patient. The clinic didn't want to run out of antibiotics and then not be able to treat someone with a bacterial infection. Because precision and recall are both important for most data science applications, the data scientists in this situation also computed *F1*, which is a harmonic mean between precision and recall. It provides a balanced metric that accounts for both how precise and robust the model is.

Representing Results
There are many potential ways to represent the results of a data science project. The two most common approaches are data visualization and modeling (usually using machine learning or statistical analysis). There are trade-offs between these two approaches. Most modeling approaches provide a quantitative summary of how well we predicted whatever it is that we formulated a problem for. Data visualization, on the other hand, offers a graphical (and often more intuitive, for most people) way of communicating the results. Of course, visualization can also go beyond being merely a representation of the results of modeling and may itself form a critical part of the data science cycle, one that works in parallel with modeling (see the section on visual analytics in chapter 4).

To represent the results of data science, that information needs to be conveyed to humans. The human visual system is the highest-bandwidth channel into the human brain (Ware 2020). We can process far more bits of information through sight (with our eyes) than we can through any of our other senses. Thus, visual representations of data are often the best ways to powerfully, effectively, and memorably convey the meaning of large amounts of data.

So much of the way we understand the world is grounded in metaphors for vision. "I see." "The lightbulb went on." "Enlightened." The visual orientation of able-bodied people has influenced Euro-Western culture to the extent that blind people can use and understand written visual metaphors as well as their sighted colleagues (Minervino et al. 2018). However, actual visualizations require explanatory text for equal access by people with visual disabilities (Asakawa 2005).

Data visualization is inherently a human-centered process. Before starting to create a visualization, you must first carefully consider your audience and your goals in communicating to that particular audience. In other words, you must consider the context rather than simply plugging your data into a tool and letting it spit out a pie chart or bar graph and saying you're done.

The first consideration to keep in mind is whether your goal is to communicate directly to your audience or to provide a visualization for your audience to explore the data themselves to inform further analyses. If your goal is to communicate, start first by understanding your audience, their context, their background, and any sociocultural considerations. Next, be very clear about what you want your audience to learn. Exactly what is your message?

Only after you've considered your audience and their context, and have articulated what your message is, should you consider creating a visualization. Now you need to think: What is the best way to use the data I have to convey my message? Sometimes the best way to communicate is simply by writing a number. A visualization may not even be necessary. At other times, the data is so vast or complicated that you need to create some type of simplification. Statistical analysis is one form of simplification, but it may not convey all the information you need or in the most effective way.

In 1973, Frank Anscombe, a statistician seeking to demonstrate the power of graphs and other visual tools to surmount the limitations of his own field, published a set of four example datasets that were very different from one another but shared not only the same mean and variance but the same regression line, the same correlation coefficient, and identical values for most other summary statistics (Anscombe 1973). In short, traditional statistics failed to elucidate the differences between these particular examples. So how, Anscombe asked, can we distinguish between such very different datasets? The answer, in this case, lay in a visualization: a simple graph, also known as a scatterplot (see figure 2.3).

Finally, selecting the proper type of visualization is critical. This is the process of visual encoding or mapping data variables to visual attributes. To do this, you must consider the effectiveness, expressiveness, and ethics of your visual choices. We will discuss this more in chapter 4. Again, it is important to provide an alternate textual description of your outcome—suitable for a screen reader—for people who may not be able to perceive a visualization (Asakawa 2005).

- Think of a project in which you did technical work and then presented your work to an audience (your class, your professor, your client, your family). Did you think about your audience as you were planning your presentation? If you had an initial presentation, how did you need to change it for each audience? What did your audience preparation teach you about your project when you tried to see it through your audience's eyes? How could you use audience preparation as a novel way of looking at your own work?

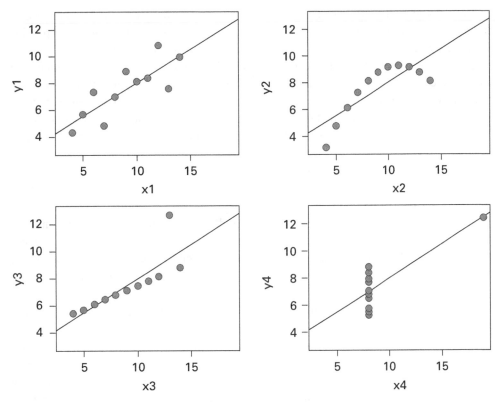

Figure 2.3
Anscombe's quartet. Four simple datasets with identical descriptive statistics but completely different characteristics (Anscombe 1973).

Distributing Results

Distributing the results is also an important part of the data science cycle. The results of the analysis depend not only on the choice of the question and the dataset but also how the concepts pertaining to the question were measured, how the data was cleaned and tidied, and all the other choices we have described so far. Therefore, distributing the results is far more than just delivering a report showing a conclusion.

What ends up mattering in a data science outcome is the complex sequence of operations in the data science pipeline, as we discuss in more detail at the end of this chapter. A pipeline contains operations for accessing data, filtering data, cleaning data, extracting features from the data, modeling, and reporting. As data science workers, we may prefer to focus our critical thinking on the model. Models seem to offer clear and principled choices: "Should we use logistic regression or a decision tree or a random forest approach?" Models are easy to evaluate, as we noted in the preceding section, using metrics such as precision, recall, and F1. For example, "The random forest approach had the highest accuracy, so we used that model." Models are also easier to "compute on," as happens with automated approaches to data science.

The simplicity of models and modeling can distract us from other aspects of data science work. This distraction is dangerous. We noted earlier that as much as 80 to 90 percent of the time and effort in a data science project does *not* involve the model but takes place during data cleaning and feature engineering. These stages are essential because they provide the model with the clean and regularized datasets it needs for computation.

Thus, what is actually *delivered* from a data science project is the pipeline, including the methods and parameters for inputting data, cleaning data, and engineering features from the data—and *then* the model. Often, the client or "user" of a data science outcome wants to receive a "product" that they can put into use in their project, service, or product. They may not want to "look inside" the pipeline to understand each step and stage. It is common for data scientists to provide a single block of code (in the form of a pipeline or a notebook or of a compiled system) that contains the detailed steps that we have described. We may like to think that precision, recall, and F1 are properties of our model. However, when we deliver the outcome to the client, it is the solution (i.e., the pipeline) that has functional properties such as precision, recall, and F1. Inadequate data cleaning and feature engineering can negatively impact those accuracy metrics. Worse, these inadequacies can also produce incorrect outcomes that can harm people's fates in medical procedures, banking systems, or criminal justice systems. Thus, in this book, we focus on the pipeline as the client-oriented outcome of data science.

Pipelines are especially easy to share in the form of Jupyter notebooks or other integrated environments. Version control systems like Git and sites that implement them such as GitHub are excellent repositories for sharing pipelines, because they keep track of all the changes in the pipeline over time. They also easily allow others to create their own copies of the pipeline. This is especially important for *reproducibility*—the idea that we need to make all the resources available and transparent so that others who are interested in related questions could reproduce our work. While traditional data science is very much concerned with reproducibility, it is especially important for human-centered data science as it concerns the people who will potentially use our results or the entire pipeline. We will discuss the importance of reproducibility in more detail and how to cultivate it at every step of the data science cycle in chapter 5.

Interpreting and Communicating Results

A key component of the data science cycle is interpreting the results and communicating them to others. We want others to be able to reproduce our results, and that requires getting out the word on what we did and how. Thoroughly documenting and describing our pipeline is important for others to be able to follow along, and especially for making explicit our mental model of how to address the questions of interest by using the dataset we chose. We discuss documentation further in chapter 3 and chapter 7.

Another important aspect of others being able to use your pipelines is the *interpretability* of your model, meaning how well the model results can be explained. In data science, this is also sometimes called *explainability*. Explainability matters because if the model is very difficult to interpret, others might not understand how it works and how to use it properly. This means that people might apply it to questions or datasets for which it is well suited, but they might apply it incorrectly and reach erroneous conclusions or use it correctly but misinterpret the results. Worse, people may apply the model to questions or datasets for which it is not well suited, thus drawing conclusions based on an invalid premise. These are all problems leading to bad data science outcomes. Hence, data scientists have a responsibility to ensure that models are as interpretable as possible. We discuss interpretability in more detail in chapter 4.

- Many people in data science tell stories about misinterpretations of their projects. Have you had this experience? What did you mean to say, and how did your client misinterpret? Can you remember the point or points in your presentation that could be interpreted in unintended ways? How could you work to reduce the likelihood of these problems in the future?

Feedback and Iteration

Much of what we've been discussing in this chapter about the data science cycle should be taken to be part of a broader tradition of feedback and iteration that runs through each part of the cycle. This is why it's called the data science *cycle*. Not only is the entire process iterative and repeatable from top to bottom, but every substep—for example, data collection or feature engineering—is iterative within itself. Moreover, feedback from one step informs a subsequent step or even prior steps in the next iteration. For example, if you decide to use a different model, you may need to clean your data in a different way, and you may need to engineer a different set of features.

Through feedback and iteration—like shifting to a different model—data scientists often explore new possibilities. At this stage, they may not want to interrupt their flow within the pipeline to document the shift. While it is completely understandable—it's hard to switch gears to a different type of task at a moment like that—this makes the pipeline much harder to decipher (both for yourself at a later time and for others). This is part of the reason we often don't know the rationale for abandoning one model and shifting to another. Such choices impede the reproducibility of results. Thus, many data scientists are now choosing to find moments of time, even in the midst of their work, where they can document, at least, their most significant decisions, such as the choice of the model. This is known as *reflective practice* and is discussed further in chapter 3.

The importance of feedback and iteration is one of the key takeaways from this chapter. In complex, social teams where many data scientists work, feedback and iteration are the norm rather than the exception (see chapter 7). The data scientist should be able to receive and *use* constructive feedback from other members of the team—be they other data scientists, domain experts, managers, or interns—and iteratively implement it within the data science cycle. To be successful at receiving and implementing feedback, data scientists need to keep a strong written narrative of this process. This documentation is usually part of Jupyter notebooks or whichever technological platform is used to construct the data science pipeline. Moreover, when pipelines are shared between data science teams within the same organization, between organizations, or even distributed to the wider world, a strong written narrative about which choices were considered from feedback, implemented, and finally included in the final pipeline is essential for the sake of transparency, reliability, validity, and reduction of future bias. The latter is especially important if data scientists return to this pipeline at a future date. Without documentation, they may forget which choices were made and how feedback affected the final production of the pipeline.

Finally, feedback is not a single-step iteration. Iteration happens multiple times and often by multiple data scientists working on the same project. Ideally, iteration over the data cycle should continue as long as it takes for the data scientist to be satisfied that the initially formulated problem has been answered rigorously and satisfactorily. In practice, time and resource constraints often prevent data science problems from being solved satisfactorily (sometimes as the result of decisions by managers or advisers with power relationships over the data scientists). This can become an impediment to the data science cycle. We discuss this issue and other related aspects more in chapters 7 and 8.

Models and Pipelines

Throughout this book, we distinguish between data science *models* focusing on the specific analytic techniques that we will cover—such as regressions, decision trees, support vector machines, random forests, topic modeling—and data science *pipelines*, which combine the model, the data collection and wrangling decisions, and the products and services that rely on it. We can usually use statistical techniques to evaluate the models, while the pipelines are frequently imported wholesale between teams or even organizations. As we mentioned earlier in this chapter, what often matters to the outcome is the data science *pipeline*. However, models are easier to evaluate using metrics such as precision, recall, and F1 (defined earlier in this chapter). This may delude us into focusing on the easy thing: working under the streetlight, as it were. In the remainder of this book, we hope to convince you of the importance of avoiding the streetlight effect when analyzing and evaluating your own use of the data science cycle.

Many data science courses teach students about constructing, evaluating, and refining models. Pipelines, when they are taught in these courses, are often represented as Jupyter notebooks or other tools for combining code, visualizations, comments, and results, or as executable graphical layouts in commercial tools such as Alteryx or SPSS. Entire pipelines are often shared on GitHub, which now supports and renders Jupyter notebooks. This means that these pipelines can be imported in their entirety by other data scientists who are interested in specific models. It is thus important to distinguish between the models, which can be statistically evaluated, and the pipelines that include those models, which are just beginning to have appropriate evaluation tools for the whole process (e.g., Bellamy et al. 2019; Olson et al. 2016).

Conclusion

In this chapter, we have outlined the standard data science cycle as a pipeline—an integrated series of steps beginning with a question and ending with an act of communication to others, such as clients. We hope that this description of the cycle can serve as an anchor point or "reference model" for the following chapters. We encourage you to begin your work by formulating a good question—even if you have to revise that question after you have more familiarity with your data, as Mao and colleagues (2019) showed in their paper on revising questions after analysis. Indeed, revision and iteration are key and expected aspects of high-quality work in data science. This is particularly true in data cleaning and data wrangling, and it is usually true in feature engineering.

The data science cycle has built-in points of self-evaluation and feedback, and these can lead to further iterations. These iterations may be based on evidence from formal evaluations, such as computing the accuracy, precision, recall, F1, and so on of the pipeline. Or these iterations may be based on your own sense of what "looks right" or "feels right" (Muller, Lange et al. 2019) as you become more familiar with your data and your tools.

Iterations are good, as long as they fit within the larger constraints of the overall project, such as semester schedules, finite grant support, and client deadlines. If iterations are good, then a flexible project plan is even better: you should anticipate iterations, and your project plan should be written around that expectation.

In chapter 3, we'll delve more deeply into some of the unseen but critically important activities that lie beneath the traditional version of the data science cycle.

Recommended Reading

McKinney, Wes. 2018. *Python for Data Analysis: Data Wrangling with Pandas, NumPy, and IPython.* Sebastopol, CA: O'Reilly. This book has broad and deep coverage of some of the messier parts of data preparation from a practical, industry-based perspective. It focuses on specific data science tools that are popular, like the Jupyter notebook, the IPython shell, NumPy, and pandas library, and includes exercises and tutorials for learning some of the finer points of data wrangling and cleaning. Some introductory programming background is expected but not required.

Rattenbury, Tye, Joseph M. Hellerstein, Jeffrey Heer, Sean Kandel, and Connor Carreras. 2017. *Principles of Data Wrangling: Practical Techniques for Data Preparation.* Sebastopol, CA: O'Reilly Media. This book explains the necessary messiness and difficulty of early stages of the data cycle.

3

Interrogating Data Science

In chapter 2, we described a standard account of how data science is done. In this chapter, we consider some of the less visible but vitally necessary aspects of data science work. This chapter's title was selected because we want to reexamine some of these standard processes from the perspective of human-centered data science. Rather than simply critique existing practices, we will look at them from new angles and make recommendations through which you can improve your own practice, ultimately making your work more useful.

One of the authors (Shion) works with a police department in a large urban area in the Midwestern United States. He is helping the crime analysis lab overcome people's personal biases when mapping crime with computer applications. Accurate crime maps, based on recent and historical crime data, result in real-life policing decisions that are made at the street level, such as how many officers are allocated to patrol a city block. He noticed that between 2013 and 2014, the state legally changed the definition of what constitutes sexual assault—yet the model that built the final maps did not reflect the updated legal definition. This changed the meaning of the maps and created a flawed policing policy of when and how officers could respond to complaints about sexual assault and whether complaints would get recorded in the police databases. There would also be disparities in any models created with this data in the future. One can easily imagine poor policy decisions being made based on erroneous increases or decreases in crime on these maps.

This story shows that we need to think very carefully about how variables change their meanings in datasets over time, and that data scientists need to remember this at all points in the data science cycle.

Data science is often presented as a clean and impersonal sequence of computational operations. Many people assume that "the human element has been scrubbed from the database," in the critique of Pine and Liboiron (2015). However, people touch the data at nearly every step. In this chapter, we will highlight how often people intervene between "the data" and "the model" (Muller, Lange et al. 2019). Data science practices often obscure these necessary human interventions that reflect choices that people have made throughout the data science cycle. If these interventions remain invisible, their invisibility may introduce weaknesses and biases into your analyses. However, transparency can often mitigate such problems.

We begin with a largely invisible step in which humans decide crucial attributes of their data science project through a *measurement plan*. Then we present five types of such human interventions (Muller, Lange et al. 2019). We emphasize that these interventions are necessary for high-quality data science work, and we hope to help you to make your own interventions in a responsible, accountable, and transparent manner:

- *Discovery* is when someone encounters a dataset and decides to use it.

- *Capture* is the process of searching for a dataset and acquiring it.

- *Curation* is the process by which data is edited or altered to make more sense to someone or is brought together from different sources.

- *Design* is how we craft or categorize data through sets of choices.

- *Creation* occurs when we or other people add values to the records in a dataset through assigning or labeling those values.

These five activities focus largely on what we *put into* a dataset. At the end of this chapter, we also consider what we might *take out of* data in order to protect people's privacy. These seem like simple activities. However, when we examine them critically, we find that there are many complexities.

Measurement Plans

A measurement plan helps people think about "what is counted and what is not, what is considered the best unit of measurement, and how different things are grouped together and 'made' into a measurable entity" (Pine and Liboiron 2015). We discussed these plans in chapter 2, but in practice they are rarely recorded. As a result, other people may see a dataset as having a "definitive" status as "the" data. In reality, the dataset was likely shaped by a series of human decisions, and possibly assumptions, that are practically impossible to account for later if they were not documented originally.

Measurement plans may not be written down because of the necessary tension between the activities of *exploring* the data and *explaining* the data (Rule, Tabard, and Hollan 2018). In exploring our projects, we try one solution and then another and then another. This iteration is a natural part of the process. We are human; we don't want to interrupt our explorations by writing explanations.

Psychologist Mihaly Csikszentmihalyi has a concept that helps explain why data scientists might not want to stop in the middle of their projects to write documentation and measurement plans: *flow*. Flow refers to absorption in the task at hand, and it is important for our ability to focus and feel satisfaction in our work (Csikszentmihalyi 1992). But there are other ways of working. Design philosopher Donald Schön argues for deliberately punctuating tasks with moments of thoughtfulness in what he terms *reflective practice*. Reflective practice helps us become aware of what we know implicitly and how we learn from experience (Schön 1984). Most of the time, we focus on the problem that we are trying to solve. Schön encourages us to take the time to focus on our own practices while we work (what he calls *reflection-in-action*) to improve our knowledge so that we can apply our now-improved knowledge to our own practices in the future (*knowing-in-action*). In Schön's analysis, this knowledge of our own practices is a necessary complement

Case Study 3.1
Documenting Data Analysis
Adam Rule, Oregon Health Science University

Data analysis is often collaborative. Take, for example, a project I helped with that looked at where text in doctors' notes comes from (e.g., is that text manually typed or copied from previous notes?). A lab-mate and I had worked on different parts of the analysis and, together, submitted our results to an academic journal. A few months later, the journal asked us to make a couple of small changes to the analysis before publishing our paper. Simple enough. Except that, in the time between submitting the paper and being asked to revise it, my lab-mate had started a PhD program at another university. I would have to replicate their part of the analysis on my own. This turned out to be more difficult than I expected, as:

- I was not familiar with my lab-mate's data or how they had cleaned it because my part of the project involved a different dataset.

- My lab-mate's analysis was written in the R programming language while I was more comfortable working in Python, so understanding their code took extra effort.

- My lab-mate's analysis was scattered across hundreds of files with names like "overall _table_stats.xlsx," "9providerstop.Rmd," and "text2.txt," which were not clearly linked.

To be fair, my lab-mate had done a number of things to make replicating their results easier—from leaving helpful comments throughout their code to using folders to separate data from code and results. Still, it took me several days to piece together the exact sequence of scripts they had used to generate their results.

This scenario repeats itself thousands of times every day in offices, classrooms, and labs around the world. It happens when a close colleague already familiar with your work wants to replicate your results, when complete strangers want to extend an analysis you've shared online, and even when you just need to rerun an analysis you wrote two days ago. Why is replicating analyses so hard? In part, because keeping a clear record of what you've done is fiendishly difficult.

Consider that "data analysis" refers to a range of activities including collecting, cleaning, modeling, and graphing data. These activities typically involve manipulating different kinds of files that take effort to name and organize (Tabard, Mackay, and Eastmond 2008; Kandel et al. 2012).

Data analyses are iterative. Rarely do you figure out what you want to do (or how to do it) on the first go. This can lead to having multiple, slightly different versions of the same dataset, script, or result, which again take effort to name and organize (Guo 2012).

Data analyses often involve branching experiments. For example, what happens if I clean the data using method A but model it using method X? Now, what if I use method B for cleaning and Y for modeling? Now B and X? Keeping track of each experiment is time-consuming and can seem unnecessary when so many of these combinations lead to "dead ends" (Kandel et al. 2012; Guo 2012).

Analyzing data is often way more fun than writing down what you're doing. What's more, annotating your code and results, in the moment, can feel like it's slowing you down.

In short, there is a tension between exploring data and explaining what you've done (Rule et al. 2018). In the end, many analyses (including my own) are tacitly "documented" through hundreds of inconsistently named files scattered across dozens of folders that can be remarkably difficult to piece back together to reproduce a result.

So how can we help analysts keep clearer records of what they have done so others can inspect, replicate, and extend their work? [Ed.: This *reproducibility* is discussed in more detail

(continued)

Case Study 3.1 (continued)

in chapters 2 and 5.] One opportunity is to design technologies that make it easier to document and organize data analyses. For example, many analysts now use computational notebooks like Jupyter or R Markdown to combine code, results, and explanatory text in a single file. While addressing part of the problem, the iterative nature of data analysis means many of these notebooks are still messy, poorly annotated, and hard to follow (Rule et al. 2018; Kery et al. 2018). A complementary approach is to establish best practices for data analysis (Wilson et al. 2017). For example, in most biology and chemistry labs, students are taught to record their experiments in a particular way (e.g., name and date at the top of a new page, list of reagents below . . .). Similar best practices are still emerging for data analysis as groups like Data Carpentry (https://datacarpentry.org/) help identify and disseminate them (Teal et al. 2015). As of yet, it is unclear which practices can be adopted from software engineering or laboratory science and which are unique to data science.

Amazing things happen when people collaborate around data. Helping them document their work in ways that make that collaboration easier remains a worthwhile challenge.

to our formal or rationality-based knowledge (Schön 2002), including the formal statistics of data science. This kind of reflection may require a change in your daily habits and take effort to achieve. Still, if you cultivate ways of working that help you think about what you are doing, and document those choices for later, there are long-term benefits.

The timelines for data science projects don't often leave room for this kind of reflection, unfortunately, which often leads to higher downstream costs in time and money. In practice, many of us stumble on or rush forward with eroding memories of exactly what we have done to get our current data and models. Even if we are working alone on a project, we may forget how the data came to be in its current shape, and how it was selected, unless we record these decisions. These kinds of records are called *provenance*. The term is derived from art history, where it refers to an "ownership trail" of a piece of art. In data science, provenance can include information about the original source of the data and also the kinds of changes that people have made to the data, such as removing outliers, transforming features, and adding "ground truth" labels to outcomes. If these kinds of changes go unrecorded, that may lead to problems for subsequent users of our datasets. Therefore, we have a suggestion: Write a memo or document to record your measurement plan, for later use by others, or even yourself. Indicate the decisions that might need to be revised or revisited. Carefully note your alternatives—the "paths not taken." This information becomes part of the *provenance* of the dataset (i.e., "Where did the data come from? What was done to the data?") and also the provenance of the model and pipeline that you are building (i.e., "Why did we clean the data in this way? Why this feature? What models did we consider before we chose this model?").

- If you received a dataset from a colleague, did you need to understand how the data had been processed or improved by that colleague? Were you able to "follow the trail" of what had been done to the data? Did you need to ask your colleague? How much did they remember? What else did you need to do?

You or your colleagues will find this document important if you need to revise your analyses or if you need to write a formal description of your work for publication or for audit.

Discovery and Capture

Digital humanist Hélène Bilis makes a distinction between data as "discovered," or donné, versus data as "captured," or capta (Bilis 2018). The *discovery* of data seems obvious and fortuitous: we find a dataset that can answer our question. However, there are hidden human actions even in discovery. *Someone* has to recognize that there is an appropriate dataset, and that same someone (or their colleague) has to decide what types of data are in the dataset and how they can be used. Thus, discovery becomes a human process, guided by human knowledge, ingenuity, and sometimes intuition (Muller, Lange et al. 2019).

Meanwhile, *capture* of data involves more assertive action, including a deliberate search for the data and a deliberate series of operations to put the data into a dataset or database. Again, *someone* understands a question or a problem statement, and that someone (or a colleague) searches for a useful dataset. Capture, too, becomes a human activity.

Where Does Data Come From (Provenance in Detail)?

To understand these concerns, you need to think through where data comes from, who creates it, and the sociotechnical processes by which data enters and flows through the data science pipeline. Data comes from many different sources and is shaped by the practices of (a) what data is easily available and can be collected relatively easily (e.g., legal availability) and (b) the actual mechanics and practices of the type of data collection process. This has implications for both biases and privacy. As datasets are cleaned, wrangled, and nudged down the data science pipeline, the definitions around the dataset can shift. Refining *and redefining* what the data represents occurs in every stage of the data science pipeline. For example, data science teams often discover that their initial question was not appropriate—or even possible—and then repurpose their data to answer a different question (Mao et al. 2019). In practice, repurposing data is not necessarily problematic, but without clear documentation of the data's provenance (Huang 2018; Stoyanovich et al. 2017) and the choices and decisions made, it is hard to fit it into another purpose.

Reusing Discovered Datasets

In some cases, it is possible to reuse an existing dataset. Researchers speak of *appropriation* when someone designs a technology or a resource for one purpose and someone else uses that technology or resource for a different purpose (Carroll 2004; Dix 2007; Dourish 2001; Krischkowsky et al. 2015).

Appropriations often are very helpful. A seemingly simple example from the real world is the curb cut—a ramped section of the division between sidewalk and street that allows people in wheelchairs to navigate more easily. Curb cuts turn out to be useful as well for people with baby carriages, people who are moving heavy objects on wheeled carts, and people using skateboards and rollerblades—a phenomenon now known as the "curb cut effect" (Treviranus 2014). People appropriate the curb cut for new usages or for new groups of users.

It may seem like a good idea to appropriate entire datasets for data science. However, this approach carries risks that may not be obvious. Suppose a dataset of streets and addresses was created as part of a real estate analysis. This dataset might be filtered by many factors, possibly including properties for sale, properties for rental, or properties with clear title

Case Study 3.2
Qualitative Research as a Data Science Skill
Anissa Tanweer, University of Washington

As an ethnographer who studies the practice and culture of data science, I have observed that often the hardest part of data science is not designing algorithms for optimization—it is figuring out *what* to optimize. One project I've studied that illustrates this is called AccessMap. In my role as ethnographer, I observed the work of the team, but I did not participate as a core member of the team.

AccessMap is a navigation application similar to popular technologies like Google Maps or Waze, except that it is designed for people with mobility limitations. Instead of assuming that someone's highest priority is getting from one location to another as quickly as possible, AccessMap makes route recommendations based on the customized criteria of its users. For example, an individual who uses a wheelchair might prefer to take a slightly longer route in order to avoid certain risks like very steep hills, intersections without pedestrian signals, or curbs that don't have ramps.

To the team developing AccessMap—a group of researchers at the University of Washington's Taskar Center for Accessible Technology—designing a new routing algorithm that optimizes for these things was the easy part. A much more difficult task was deciding how the underlying data about the pedestrian network should be generated, represented, and treated.

For example, the team had a lengthy conversation about how to account for "street furniture" like benches in a cost function for AccessMap routing. If a bench narrows the sidewalk, perhaps it should be labeled as an *obstruction* to someone in a wheelchair; however, to someone walking with a cane who needs a brief rest, perhaps it is an important *amenity*? The act of encoding these priorities forced the team to be explicit about what is valued by different stakeholder communities.

The team members realized that every choice they made involved different kinds of values and had different consequences for different communities—enabling some uses while inhibiting others, elevating some perspectives while suppressing others. But how did they know what values were important to different people, and how did they go about weighing these trade-offs? After doing a stakeholder mapping exercise informed by leadership's extensive experience working with communities of people with disabilities, the team did a lot of qualitative research among a range of diverse constituencies. This included interviewing different stakeholders, reviewing archival materials, and conducting fieldwork.

When deciding how to represent curb ramps in AccessMap, the team entertained two extremes. They could visually represent curb ramps as polygons on the map, which would provide relevant information about *each* ramp's exact location and dimensions. But mapping these would be difficult because it required a high degree of precision. Or they could simply place a node (or dot) on a corner to indicate if *any* curb ramps were present regardless of how many and where exactly they were. Mapping would be much easier using this method, but it would result in a loss of information.

The importance of these choices became clear to the team members one day during a site visit where they learned how a local public transportation agency evaluates accessibility of the built environment. The AccessMap team walked around the city with contractors from the agency, carefully examining sidewalks, curbs, and crosswalks while discussing what made those features accessible or inaccessible. At one point, they stopped to contemplate a corner where there was only one curb ramp that was physically distanced from two crosswalks leading into the intersection. This scene prompted one of the team members to reflect on something they had learned several weeks before, while interviewing wheelchair users.

The team member recalled that different interviewees seemed to exhibit different levels of "adventurousness." One set would be okay with this situation—as long as there is a curb cutout *somewhere* along the block so they can get off the sidewalk without toppling over,

Case Study 3.2 (continued)

they don't mind traveling in the street for short distances. Other people, however, need the curb ramps to be "optimally placed on crosswalks" so they don't end up traveling in the street among traffic where motorists would not expect to see them. So, for some people, it would be helpful to see *precisely* where the curb ramps are located because, in a case like this, they might not be comfortable with how far the curb ramp is from the crosswalks. This made the team realize that using a single node to indicate whether or not any curb ramps exist on the corner would probably be inadequate for some of their target users, and that it would be important to map the location of each ramp. They took this into consideration alongside other things they learned from the open-source mapping community while attending meetup groups and studying discussion board records—namely, that requiring too much precision would inhibit people from adding curb ramp data to the map at all. The team ultimately ruled out both of the initial ideas—representing each curb ramp as an individual polygon and representing multiple curb ramps with a single node. Instead, they compromised between the two, deciding that each unique curb ramp should be mapped as its own node positioned as precisely as possible. This path forward came into focus for the team while closely observing the real-world conditions their data represented and combining that experience with insights gleaned from interviews and archival research.

AccessMap involved a lot of work that many would recognize as "standard" data science tasks, including the synthesis, cleaning, and algorithmic processing of large, heterogeneous datasets. But the team also relied on many layers of qualitative research activities that are often overlooked as critical parts of data science practice.

To read more on the deliberations undertaken by the AccessMap team, see Tanweer et al. (2017). To view AccessMap, see: https://accessmap.io.

(established ownership). In an analysis of real estate property values, there might be a lower limit for value, and properties below that value might be excluded. Properties with no recently established value might also be excluded as a form of missing data. Properties without a clear title might be excluded on the basis of a different type of missing data.

Now suppose that you want to reuse that dataset for a census-like analysis of family structures in a neighborhood. If your focus is on families rather than on property values, then the filtering that shaped the dataset might cause you to omit some families from your analysis because they live in buildings that were excluded by the filter. It also means that the filter might have excluded low-value properties with unclear titles. In this case, the process of filtering introduces a bias into the dataset because it would tend to include wealthier families, on average. Because of income inequality and wealth inequality across cultures, races, and immigration status in many countries, the resulting dataset may underrepresent people from particular groups. This is a good example of why you should record the measurement plan. If the history of the dataset is neither indicated nor accounted for, then someone working with it might not know that it has become—for these appropriated purposes—a biased sample, leading to biased analyses and biased outcomes.

Discovered data can also lead quite innocently to a biased dataset, with negative impacts on people based on demographic categories. This is why one of the first steps of working with discovered data is finding out its provenance, including its history, its original purpose, and how it may have changed over time—sometimes also called "data lineage" (Mayernik et al. 2013). Provenance may matter even in data that appears to have nothing to do with humans. When one of the authors (Cecilia) was developing an astrophysics database, one of the astronomers told her to make sure her new database included not only stellar classifications but also the *name* of the individual astronomer who originally assigned that classification. Apparently, one senior scientist tended to classify stars in a manner that many of the others in the group disagreed with. Thus, even data that appear to be "objective" can contain human bias. This leads to our next recommendation in this chapter: When you reuse datasets, check the history of the dataset to learn how it may have been processed in a way that could affect your analysis. These changes might distort your analysis or your outcome, but you may be able to reverse them. Or you may need to weigh the costs and benefits of working with a dataset that may be systematically incomplete or unintentionally biased. A dataset without history—without provenance—may *not* be safe to use in some cases.

- How do you record the series of transformations that you make on a dataset? Do you rely on the hope that people will read and comprehend your source code? Do you create in-line comments? Do you write a separate document? If you are working in a Jupyter notebook or another literate-programming environment, do you take advantage of the documentation resources within that environment (e.g., markdown cells in Jupyter notebooks)?

- Who is likely to be the "next user" of your datasets and analytic code? Do you think your documentation practices will be sufficient for their needs? Why or why not?

We would like to think that most people with data science projects keep a clear history of the series of necessary and responsible computations and transformations that they perform as they work on a dataset or a model. Unfortunately, in practice the available data tells a different story.

It is quite common for data scientists to use programming environments such as Jupyter notebooks. These online environments are designed so that "programmers write code and explanations for use by other programmers or by other stakeholders" (Knuth 1984). The basic structure of a Jupyter notebook is a series of windows called cells. In the original version, the cells could be of two types: *code cells* containing executable source code, and *markdown cells* containing formatted text to explain or comment on the code cells.

In a study of one million publicly shared Jupyter notebooks, researchers discovered that the markdown (or documentation) cells were severely underutilized, and many notebooks had no documentation cells (Rule, Tabard, and Hollan 2018). Of course, this does not mean that people are not commenting at all. In a study of fifty-two academic computational notebooks, researchers found that 82 percent included comments within the code cells. Nevertheless, these were almost exclusively used to simply describe what the code was doing. Only 10 percent of notebooks used comments to explain the analyst's reasoning or decisions, and they tended not to provide resources for the next person who will work on the data or code. This apparent shortcut can have negative consequences, as we've described. In chapter 7, we will return to documentation as a practice that supports collaboration in data science.

Choice and Effort of Datasets
There is often a trade-off between cost versus depth of analysis, where cost is typically measured in time and effort and depth is typically measured by utility and insight from the

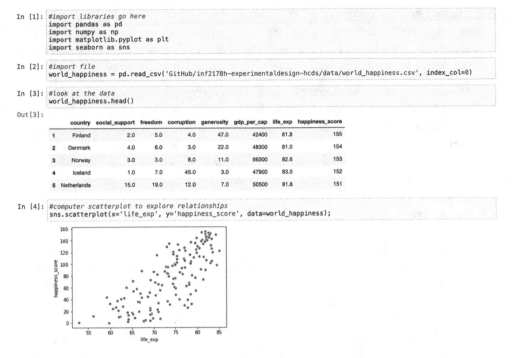

```
In [1]: #import libraries go here
        import pandas as pd
        import numpy as np
        import matplotlib.pyplot as plt
        import seaborn as sns
```

```
In [2]: #import file
        world_happiness = pd.read_csv('GitHub/inf2178h-experimentaldesign-hcds/data/world_happiness.csv', index_col=0)
```

```
In [3]: #look at the data
        world_happiness.head()
```

Out[3]:

	country	social_support	freedom	corruption	generosity	gdp_per_cap	life_exp	happiness_score
1	Finland	2.0	5.0	4.0	47.0	42400	81.8	155
2	Denmark	4.0	6.0	3.0	22.0	48300	81.0	154
3	Norway	3.0	3.0	8.0	11.0	66300	82.6	153
4	Iceland	1.0	7.0	45.0	3.0	47900	83.0	152
5	Netherlands	15.0	19.0	12.0	7.0	50500	81.8	151

```
In [4]: #computer scatterplot to explore relationships
        sns.scatterplot(x='life_exp', y='happiness_score', data=world_happiness);
```

Figure 3.1
Sample Jupyter notebook with code and comments.

data. If we have scarce resources, we may need to opt for the most convenient data in the most convenient format. Using easily acquired and well-formatted data may seem like a good engineering decision: Why should you do extra work if it does not seem necessary?

However, there may be unintended additional project costs associated with the lowest-cost data. The lowest-cost data may also be the lowest-insight data. In the long run, it may be worthwhile to do the extra work to obtain a more expensive dataset—and by expensive we mean time-consuming, effortful, or literally costly.

One of the authors of this book (Cecilia) worked for many years as a software contractor, traveling from company to company writing code for one project and then moving on. When she first started her career, she wondered if all the effort she put into documentation was worth the cost, given that she spent a relatively short time at each location. However, after a few years, she found that her former employers were recommending her to other companies because her software was easy to understand and reuse, as opposed to the all-too-common "Kleenex" or disposable code they usually had to deal with. Thus, this type of careful work provides long-term benefits, both for companies and for individual careers.

So, our third piece of advice is: Consider the short-term *and* long-term costs and benefits—to you and to your colleagues—as you choose your dataset. Evaluate the trade-offs between ease of acquisition versus depth of insight. Write a memo or document to explain your decision. In the real estate example that we described earlier, it is important for the census project to know if certain houses had been excluded from the original dataset because they had unclear title or uncertain value. For the census project, those *real estate exclusions* may also inadvertently become *racial exclusions* when the dataset is used for a new purpose. This kind of provenance information can be included in your measurement plan, in a separate document, or in a markdown cell in a Jupyter notebook.

- How do you and your team make these kinds of cost/benefit decisions about the expense in currency and labor for each alternative dataset? Data science has roots in multiple disciplines, such as math, economics, engineering, and management science. Our engineering and management science roots would encourage us to find the "best" data at the "least" cost. How do you and your team determine what is "best," and what near-term and long-term costs do you consider? How do you make these decisions?

Curation of Data

The next step after data have been put into a dataset is *curation*: Which records are useful? Which should be excluded from analysis? We call these decisions that people make about data a process of *curation* because it entails creating and structuring a collection of things (records, in our case) and deciding which things to include into a collection (a dataset), like a curator does for objects in a museum or gallery exhibition. These decisions shape the data in many ways and should be recorded as part of the dataset's provenance. It may seem that these decisions are part of the measurement plan (see above), but often these decisions are made *in the moment*, weeks or months after the measurement plan has been completed (Muller, Lange et al. 2019).

This curation process also happens when we collaborate on datasets with others. Collaborators curate the views of data offered to their collaborators to highlight particular things that are important from their perspective (Taylor et al. 2015). For example, telemedicine sometimes involves work of "crafting the image" so that the remote doctor can understand the structure of the image-based data (Mentis, Rahim, and Theodore 2016). Mars rover visualizations often entail curating images to emphasize characteristics in data useful for particular groups of scientists (Vertesi 2015).

Our fourth recommendation to you is: Understand who controls how others can see the data in a data science dataset. What patterns of collaboration, shared views, and *prepared* views are taking place? What aspects of the data do you need to make visible, and to whom?

- When you choose or exclude data—or shape data—what practices and guarantees of quality do you use? Can your "next user" understand what you did? Will your preparation and selection of data and related details meet their needs? How can you tell?

Taking everything together, our fifth suggestion should come as no surprise: Write a memo or a document to record your understanding of curation for this dataset, both to clarify your decisions for the future and to track changing needs over time.

Design of Data

Chapter 2 described data cleaning and wrangling, which are necessary to prepare a dataset for analysis. Now, we want to reexamine those concepts as human activities.

There are many cases of scientific researchers creatively categorizing their data to make it suitable for analysis (Bowker and Star, 2000). They might, as an example, include the use of established color charts to describe the appearance of earth samples; the description of biologically and minerally complex lake samples as a limited and manageable set of attributes; and the classification of species according to previously established taxonomies (Feinberg 2017). We do this kind of task whenever we prepare a dataset for analysis. The goal is often to reduce the diversity of our messy data into seemingly straightforward classes and subclasses, represented by labels in one or more columns in a database.

This is called the design of data, and it takes many forms. People "design" data whenever they see it through various disciplinary conventions that help us to turn a rich and complex world into a manageable number of categories (Bowker and Star 1998; Feinberg 2017). It is not that design of data is separate from data analysis, but that it is integrated into how we *make* data in the first place (Star 2002).

Your task as a data scientist is to *craft* an interpretation that will be useful for analysis and to record that interpretation as a class or subclass of data. The act of crafting implies human intention, human action, and in many cases human creativity. These acts of crafting have *consequences*. Consider the experience of airport security screening, as Sasha Costanza-Chock (2020) has written about. An algorithm that has been designed in terms of "male" or "female" bodies will flag a trans woman for further screening by a security officer because she does not conform to a system designed for binary gender. In this way, the designed features of "male" and "female" in the scanning system lead, for trans travelers, to systematic and repeated delays, possible missed flights, and of course discussions of intimate personal details and histories with strangers. One proposal to reduce these kinds of problems is a broader characterization of gender identity, XM (see DeCamp 2020).

Data Cleaning

Cleaning data is one way in which we design data. When we clean a dataset, we may think first about statistical distributions. Many modeling algorithms require quantitative data to be as close to normally distributed as possible (Osborne 2008; see chapter 4 in this book for a more detailed discussion of data distributions). Qualitative data—such as text, video, and images—may be affected by decisions about the categories that we use to code qualitative data into categories for analysis. Cleaning, however, also entails other considerations.

Suppose you are trying to exclude outliers, a common practice in preparing a dataset for analysis. How do you *decide* the criteria that determine whether a record contains outlier values? In our previous real estate example, we looked at the impact that decisions for outlier values in property values might have on the demographic representativeness of families. Do your outlier criteria tend to exclude people of a particular class, race, or gender? If the answer is yes, then you may wish to reconsider your outlier criteria. "Outliers" may actually define a class or a subclass that could be important to your analysis—or to the impact your analysis has on other people. Outlier values may show that you should include more classes or subclasses in your data. For example, outliers could be an indication that what you are studying is more complicated than you originally thought, and that you can potentially describe more of that complexity in a more precise, accurate, and

Case Study 3.3
Pizza Data Design
Melanie Feinberg, University of North Carolina

Imagine that you've been given the following project for your class: to connect the hungry people in your town with the best pizza for their circumstances. To accomplish this task, your team will design a dataset to describe local pizza purveyors. What kinds of data do you need, and how will you collect it?

Design a dataset? We tend to think of data as established facts. It doesn't seem much like design just to assemble facts. But another way to perceive data is how we *make* facts—the facts that we need to accomplish a particular purpose. In terms of your pizza project, if users are looking for "the best" pizza, what characteristics of pizza should you consider? As you discuss this with your team, you find that everyone focuses on different properties. One team member announces that crust thickness is the most important attribute of pizza. Another responds that texture matters more than thickness, while a third suggests that maybe both need to be considered together. Someone proposes that the cheese (type and amount) contributes more to the pizza than its crust; someone else chimes in that toppings (preparation, combinations) are what make a pizza really distinctive. Pondering all these elements, one teammate wonders about comparative quality. How do we differentiate between a chain pizza and a gourmet pizza? Ingredients? Type of oven? Are ratings necessary? Someone else disagrees vehemently with this suggestion. Isn't quality subjective? Data shouldn't be subjective. And anyway, what about price? Or the open hours? The best pizza is the one that I can get for the least money at midnight on Saturday, that person contends. Suddenly, what you thought would be a ten-minute activity turns into an intense debate. After a few hours, your team is tired of arguing. You decide to collect data for three pizza attributes that everyone agrees are important to know: the restaurant location; the price of a medium, two-topping pizza; and the thickness of the crust. In order to ensure that each of these attributes can be objectively determined, you decide to measure the crust thickness in millimeters.

As part of your project instructions, each team member is required to independently collect data for a test case, University Pizza. You aren't quite sure why you need to do this; once you've designed the schema, collecting data seems like a brainless task. But when your team compares results, everyone's data is different. For instance, University Pizza has three locations in your town. You listed the addresses of all three. But your classmates each chose a single location—different ones, of course. Only now do you realize that "pizza purveyor" can have multiple interpretations: it could be a *business*, in which case all locations would be listed together, or it could be a *branch*, in which case each location would be listed separately. Price data also differed. One person went to collect data on a Wednesday, when a medium pizza is discounted by a dollar. One person used the price from the website, but that information was out of date. One person included tax; another person rounded to the nearest dollar. All your data is like this, full of variation. What happened?

As an instructor, I've seen this situation many times. The variation in your data is normal. It results from the human process of *data design*. Initially, students imagine that data emerges from the world without much human intervention. But when we start to think about a particular situation—even something as mundane as obtaining pizza—we realize that we have multiple possibilities to express that situation in data. When selecting between these design possibilities, students often believe that countable data will be more reliable. But countable data can be just as ambiguous as more "subjective" assessments. Moreover, the decision to prefer quantitative measurements—which is itself a subjective choice—can affect data utility. (A human judgment of thin, thick, or medium crust will probably serve our purpose better than thickness in millimeters.) Likewise, students tend to believe that data collection is a mechanical process of rule following. In reality, data collection is an

Case Study 3.3 (continued)

extension of design. Many small decisions may have cumulative downstream effects. (Deciding whether a pizza purveyor is a business or a branch affects all of the subsequent data.) Determining how to aggregate data collected under different conditions (different team members, different times, different collection instruments) also involves decision making. When students perceive the inconsistencies that result from these layers of design, they often blame their inexperience. In "real life," they imagine, such uncertainties do not appear. But all data shows the marks of accumulated design decisions. That doesn't make the data terrible! It makes the data human.

powerful model—perhaps using a more strategically sampled dataset that includes a broader range of humans.

Further, outlier values may indicate that you may need to analyze *intersectional* combinations of race, class, and gender (Collins 2015; Foster and Hagan 2015). For example, transwomen of color experience discrimination differently from other women because of their intersecting experiences of race, gender identity, and socioeconomic status (Arayasirikul, Wilson, and Raymond 2017). If your outlier criteria risk excluding or underrepresenting a particular group, then you may wish to reconsider the criteria for selecting outliers to have a more complete or fair model.

A common approach is to use a binary dependent variable, asking, for example, is this a picture of a cat *or* a dog? The analysis is easiest to perform if the binary variable describes a 50–50 percent split in your data. However, some groups make up far less than 50 percent of the population. For example, Indigenous peoples in most countries are represented at very low frequencies in many datasets. If you decide to compare, for example, "white" versus "nonwhite," then the smaller Indigenous groups in your data may become statistically invisible among larger "nonwhite" groups (see Smith 2013 for a discussion of the "gaze of the researcher"—that is, taking into consideration *whose* language is used to describe a less powerful group). You may need to develop a strategy to do multiclass rather than binary testing. Or you may need to use other sophisticated statistical techniques to reduce the majority group representation in your dataset until it matches the frequency of specific other groups to allow for binary testing.

All of these inclusion/exclusion decisions are how we necessarily *design* the data and the dataset. As a consequence, we advise that while you determine your thresholds and/or your rules for dealing with outliers, it is important to test the impact of those outlier criteria on the resulting edited dataset. Do the outlier criteria tend to exclude people from particular demographic categories? Do the outlier criteria affect any other attributes that are important in the context of your project? If the answer is yes, then consider whether you can manage outliers in other ways.

- How do you decide whether a record contains an outlier? Is it your decision, or do you consult colleagues and come to a shared decision? How do you analyze-forward to test the impact of your decisions about outliers? How do you record your decisions so the "next user" of the data will understand what you did?

Data Wrangling

As we discussed in chapter 2, data wrangling often involves creating new features based on combinations of existing features (Anderson and Cafarella 2016). Each of these "engineered features" is itself an exercise in design.

Let's look at a case involving the colleagues of one of the authors (Michael). In a financial modeling project, a data scientist wanted to engineer a feature that could summarize each person's employment history. The idea was to compute a "percent worked" index for each person, as one of several predictors. The intuition was that someone with a stronger employment history would probably be more likely to repay their loan successfully. The data scientist's first attempt to engineer the feature was to take the number of weeks that the person had worked and divide it by the number of weeks in the person's life. This computation created a simple "percent of weeks worked." What could be wrong with this ratio?

It turned out that there were a number of problems with the index. The first was that it was biased against young people. Some nineteen-year-olds may experience full employment for the first time at the end of secondary school education. The percent-worked ratio for a person of this age could be as low as 1/19, or approximately 5 percent. By contrast, a thirty-eight-year-old from the same country could have a ratio of 19/38, or approximately 50 percent. If this ratio were used as part of a loan application process, then older people would be more likely to qualify for loans and younger people would face an unintended *structural barrier* to loan qualification based solely on the way of computing the "percent worked" index.

And there were further issues. In many cultures, women take more time away from work to care for children and elders, thus incurring a penalty in the "percent worked" index. Also, people from marginalized groups may face employment discrimination, resulting in fewer "weeks worked" than their nonmarginalized peers.

This data scientist had no intention to create a structurally discriminatory index. What could they have done? One possibility was to recalculate the "percent worked" index, based on a different starting year for the computation—for example, at the end of secondary education. This change would remove the age-based impact, but it would do nothing for gender-based or race-based impacts. Realistically, there are other, more complicated and technically sophisticated ways to deal with this issue, and sometimes it is hard to admit that there is no accounting for every issue. Sometimes there may be no choice but to simply omit an inherently biased feature. You would not want to use race as a predictor for bank loans. Based on this discussion, you might not want to use "percent worked" as a predictor for the very reason that it disadvantages people from marginalized groups. It becomes an indirect (and often unintentional) code for that marginalization.

And yet the data science worker in this story was unaware of the fundamental problems with their "percent worked" index. The engineered feature was about to become part of the data wrangling aspects of the pipeline, where its disproportionate impacts would have become invisible to anyone who worked with the model later. Fortunately, in this case, the feature was removed from the final product. The lesson here is that choices in feature engineering and data wrangling can have serious consequences for the accuracy and fairness of our models and their outputs.

We advise: While designing features, consider which features may have a negative or adverse impact on groups of people. Are there alternative ways to compute your features that have less impact? These are difficult issues to think about, and we recommend working on them with people you trust.

• How do you examine your features for their unintended impacts? Do you do this by yourself, or do you involve colleagues? If you are concerned about the impact on particular populations, are you able to consult with those populations to understand their perspective on your data work? How easily can you change your feature designs as a result?

Ways of Thinking about the Design of Data

Designing data deals with how we prepare data for modeling. The design of algorithms and the design of data are areas in which personal or societal values may enter into our computations, for good or ill. We showed several examples of how seemingly straightforward operations could have unintended consequences or build prejudices into a project that was intended to be fair and unbiased. Researchers have argued that one way to mitigate potential problems is to include the stakeholders' values in the design of any project, data science or otherwise. Two such frameworks for doing this are known as *Value Sensitive Design* (Friedman and Hendry 2019) and *Values in Design* (Flanagan and Nissenbaum 2014).

We can also describe data as designed through the data science process. There are important consequences of this perspective. Let's reconsider the real estate database that we described previously. We can think of the real estate database as being designed by people working in data science, and we can begin to show the actions of humans in that database when we populate the database with its human stakeholders (Muller 2011). The act of designing is usually a matter of "designing for," and so we can ask "designed for what?" and "designed for whom?" For full transparency and accountability we can ask "designed *by whom*?" In an organizational setting, we may also need to ask, "Whose rules or policies affect the designer's choices?" The realization that people design data may make us wary, because we may hold on to the notion that data reflects an objective reality about the world. But asking what other influences may be shaping the data, model, and pipeline will help us design better data science projects.

If you are reusing a dataset, we have more advice: Attempt to find the history (provenance) of the dataset. If you can contact the creator, ask them to explain their decisions about records and their inclusion and exclusion criteria, and how they created any engineered features.

Creation of Data

Finally, we consider what happens if some data fields are not available. A crucial topic here involves *ground truth* (first discussed in chapter 2). In a typical supervised learning project, the data science pipeline computes a predicted value, such as a classification or a number, for each record in the dataset. When we train a data science model, we input a subset of the data, known as the training set, that includes values or labels based on empirical observation on each record in the subset. The values or labels that we are trying to predict have been named "ground truth," although "truth" may be a misnomer in this case. The model then "learns" how to predict the ground truth for each record in the dataset, based on the predictors or "features" in each record in the training set. The ground truth is usually considered to be the definitive information about the outcome— that is, the definitive classification or the definitive numerical outcome. For example, we might want to predict a medical diagnosis. We would use medical test results as our predictors, and we would use a doctor's diagnosis as the ground truth.

However, sometimes the ground truth value is not available in the dataset, and there is nonetheless a societal or organizational need to be able to create a data science pipeline to make predictions. Research has shown that in some cases, one or more members of data science teams had to create the ground truth data (Muller, Lange et al. 2019). While some informants in this study considered ground truth to be an "objective fact," others evaluated as many as three different data sources to find a ground truth that could be captured and analyzed at scale. Some study participants acknowledged that their available ground truth was really a matter of opinions by domain experts. Other informants stated that, in the absence of any other source of ground truth, they had initiated what they believed was a high-quality process to create the ground truth on each record in the dataset. Two informants said, "I am the ground truth" (Muller, Lange et al. 2019, 9).

As in the case of designed data, we are far from the idea of data that has been scrubbed of human influence. These ground truth values may be *entirely* constructed by one or a very small number of humans (Muller, Lange et al. 2019). If we lose sight of that influence, then we can fool ourselves about the "objectivity" of a data science analysis. As above, we encourage you to think critically about the source of your data, find out how the ground truth values were obtained or created, and record your own actions to define and create (if necessary) ground truth values in your dataset.

As a data scientist, you must be aware of people's fundamental influence over data and datasets. Data does not necessarily have a kind of authority derived from some separate existence or status of the data. People leave their fingerprints all over the data, beginning with how they initially acquire the data through discovery or capture and how they manage the data prior to analysis. Human influence on the data continues through data curation and design. By taking these human realities into account, we can improve the quality of our data science work and provide higher-quality resources to our colleagues.

Privacy and Reidentification

So far, we have written mostly about what we *add* to a dataset when we curate, design, and create the data. Data science workers sometimes also work hard to *remove* certain types of data from a dataset. This effort is often more complicated than it would initially seem.

In the mid-1990s, Massachusetts Group Insurance Commission (GIC) decided to release anonymized data on state employees, including every single hospital visit. The motivation was to help researchers. To ensure the privacy of state employees, the dataset was stripped of obvious identifying information, such as names, addresses, and Social Security numbers. The governor of Massachusetts at the time, William Weld, assured the public that patient privacy would be protected by removing these identifiers. In response, then-graduate student Latanya Sweeney decided to make a point about the dangers of reidentification (Sweeney 2000). She knew that the governor lived in Cambridge—a city of just over 50,000 residents with seven zip codes. For only $20, she obtained the voter rolls for the city of Cambridge. This second dataset contained the following information for every Cambridge voter: name, address, sex, birthday, and zip code. Combining the two datasets, Sweeney easily identified Governor Weld. Only six Cambridge residents shared Weld's birthday, only three of them were men, and only one man lived in his zip code. To make her point, Sweeney sent the governor's hospital records—complete with diagnoses and prescriptions—to his office. To illustrate that the governor was not alone in this vulnerability to reidentification from supposedly anonymous datasets, Sweeney further showed that 87 percent of the US population could be uniquely identified based only on zip code, gender, and date of birth.

Privacy is more complicated than it appears at first glance, and it is defined and imagined in many different ways by industry, academia, and the ordinary person, as well as across cultures (Fiesler and Proferes 2018; González et al. 2019). In this section, we look at privacy as an issue when considering and working with datasets (see "Computational Privacy Tools"). In data science we may be given large amounts of sensitive yet anonymized data. As the example above and countless others have shown, such data often can be reidentified with relative ease. It falls to data scientists to consider these issues and take steps to mitigate and minimize the potential privacy risks in the datasets we work with. We can also report privacy concerns within our organizations, inform decisions and policies to protect privacy, and advocate for better regulations to prevent future occurrences.

Privacy and Consent: Who Counts or Who Makes It into the Dataset?

A part of interrogation of the data science pipeline, especially around datasets that deal with individual and social data, also needs to be around who makes it into a dataset. Are people put into datasets with their informed consent? Or are they swept up along with everyone else? We note that, in certain regions, these questions have legal implications (see Voigt and Von dem Bussche 2017).

Here, we will imagine privacy as a lack of anonymity or identification of specific people in our datasets, as that is most relevant to the data science pipeline. This is not the only way to imagine privacy, however. In fact, there's been plenty of work done that unpacks how different people think about privacy (Terzi et al. 2015; Yu 2016; Fiesler and Proferes 2018; González et al. 2019). We acknowledge this here but focus on how professional data scientists must grapple with privacy implications in the process of doing work in the data science pipeline.

Data scientists should consider these issues iteratively during problem formulation, data collection, and the data cleaning/wrangling phases. Overt and covert decisions are made during these steps in the pipeline that could result in potential privacy violations if anyone were to access so-called anonymized datasets. In fact, evidence suggests that no dataset that contains human data is perfectly anonymous, so we can only work toward minimizing potential future privacy violations and other issues.

Issues in Combining Datasets

Multiple issues occur when datasets are combined. For instance, individual datasets might do a great job in protecting the privacy of people whose data is being analyzed, but when datasets are combined, these privacy protections are lessened. As Latanya Sweeney's work showed, combining US zip-code data with practically any kind of individual or social dataset brings potential privacy violations.

Legal and regulatory frameworks should also be considered. For instance, in the United States, the Health Insurance Portability and Accountability Act (HIPAA) and Family Educational Rights and Privacy Act (FERPA) regulations apply to electronic health records. To prevent violations of legal protections of privacy, data scientists must consider legal requirements when cleaning, wrangling, transforming, or combining datasets. The same holds true in other countries and jurisdictions, such as the European Union's General Data Protection Regulation (GDPR) and California's Consumer Privacy Act (CCPA).

Computational Privacy Tools in the Data Science Pipeline

To address some of these issues more computationally, advances have been made in the past decade or so to attempt to protect privacy by either (a) quantitatively minimizing the amount of information that is used to model useful predictive outcomes or (b) engineering features in a more useful and efficient way that doesn't need to maximize information to be useful for modeling. While we introduce these privacy-preserving methods here, they are rarely covered in undergraduate or even professional graduate data science programs. There are plenty of resources for more information if you want to preserve privacy in practice. We will outline a couple of popular approaches here.

- *Differential privacy* is a system for publicly sharing information about a dataset by describing the patterns of groups within the dataset while withholding information about individuals. The underlying idea is that an observer seeing the output of a model cannot tell if a particular individual's information was used in the computation. A more technical way to describe differential privacy is that it constrains the algorithms used to publish aggregate information about a statistical database, which limits the disclosure of private information of records in the database. In recent years, differential privacy has found popularity in many industries that have data science applications. For example, differentially private algorithms are used by government agencies like the US Census Bureau to publish demographic information or other statistical aggregates while strengthening the confidentiality of survey responses. Companies like Apple use these algorithms to collect information about user behavior while controlling what is visible even to internal analysts.

- *K-anonymity* is a property of a dataset, usually used to describe the level of anonymity in a dataset. A dataset is k-anonymous if every combination of identity-revealing characteristics occurs in at least *k* different rows of the dataset. This is illustrated in figure 3.2. Many data science applications consider k-anonymity, particularly in cases where there's a combination of several different demographic variables that could potentially lead to reidentification.

These computational tools for preserving privacy in datasets are not magic wands that eliminate privacy concerns. They still require proper evaluation of potential outcomes. Many of these techniques are mathematically sophisticated and difficult to apply in practice without highly specialized training. However, you should know that these tools exist, and you can seek additional training on techniques to help preserve privacy in the datasets that you analyze.

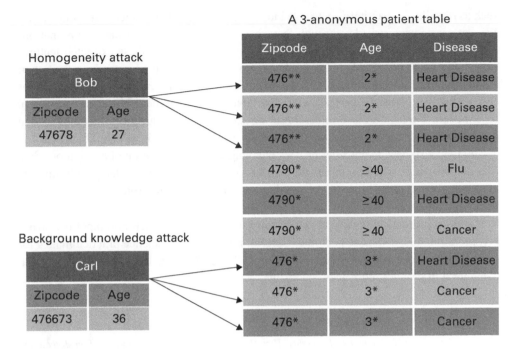

A 3-anonymous patient table

Zipcode	Age	Disease
476**	2*	Heart Disease
476**	2*	Heart Disease
476**	2*	Heart Disease
4790*	≥40	Flu
4790*	≥40	Heart Disease
4790*	≥40	Cancer
476*	3*	Heart Disease
476*	3*	Cancer
476*	3*	Cancer

Homogeneity attack

Bob

Zipcode	Age
47678	27

Background knowledge attack

Carl

Zipcode	Age
476673	36

Figure 3.2
K-anonymity example from patient and zip-code data.

Conclusion

With the information in this chapter, we've introduced some of the complexities to think about as you work within the data science cycle. To begin, we advise that you focus on the measurement plan, which defines the project and provides key insights into how data science teams are approaching the problem to be solved. As we noted, measurement plans may not be documented, and many of the human decisions made when working with data also may not be documented. The absence of documentation can interfere with proper interpretation and can confuse or misinform colleagues who subsequently work with the data or the code that manipulates the data. We recommend that you keep careful notes, in the form of memos or documents, so that you can remember what you have done with the data and so that your colleagues can make sense of the data if they work on the project after you. We encourage you to cultivate the ability to reflect on your work at the same time you are engaged in it.

Recall the five ways that people intervene between "the data" and "the model." Data *discovery* requires someone to identify relevant datasets and entails human judgment. Data *capture* relies on judgment when someone searches for a dataset, and then reformats it for use in a data science environment. Choosing among different records and excluding some because they do not include valid values is called data *curation*. Curation also happens when someone decides that the data includes outlier values. However, we caution that an outlier value may indicate that there are some records that are *systematically* different from the others. Data *design* takes place when data science workers adjust the data, such as recoding a quantitative variable into classes such as "low," "medium," and "high." We can create new variables through feature engineering, using nonlinear combinations of existing variables such as ratios of existing variables. Each of these extracted features comes from choices people make and may intentionally or inadvertently reflect their assumptions or biases. Finally, data *creation* occurs when people make their own decisions about what they think the true value of a variable is, or when they solicit the opinions of other people to establish those values. A strong and relatively frequent type of data creation occurs when humans must provide values for the ground truth variable on each record.

You may also need to consider privacy challenges. Reidentification of anonymous data is a thorny issue. When anonymized datasets are combined, particular people may be identified from them. This leads to ongoing questions of how to protect people's privacy in large-scale data.

If you keep these challenges in mind, you can make your models and pipelines stronger, more reliable, and more accurate. Careful, reflective data science practice makes for more ethical data science—which is also simply better data science, because your ethical behavior will produce more accurate results.

Recommended Reading

Eubanks, Virginia. 2018. *Automating Inequality: How High-Tech Tools Profile, Police, and Punish the Poor.* New York: St. Martin's Publishing Group. This book has a gripping set of stories of how technology affects human rights.

O'Neil, Cathy. 2016. *Weapons of Math Destruction: How Big Data Increases Inequality and Threatens Democracy.* New York: Crown. A critique of data science "certainty," written for the general public.

4

Techniques and Tools for Data Science Models

In this chapter, we examine several common types of data science models, show examples of each, then discuss the rise of tools that automate many of these models and processes, like data cleaning and wrangling and bias mitigation and detection. We start by describing machine learning models stemming primarily from computer science research, statistics models stemming primarily from statistics research, automated artificial intelligence (AI)/machine learning models primarily from industry research, and automated or semi-automated models for bias detection and mitigation. We also look at the role of decision making in all of these models. Our goal is to give an overview of each kind of model. Other texts can provide in-depth detail on how to use them. Here, we summarize some of the key advantages and challenges so that you can compare them. Finally, we conclude by returning to our discussion of visualization (first mentioned in chapter 2). It is an important technique that can be used to represent and communicate data science models and as part of the data science analysis process itself.

Machine Learning Models

In the broadest sense, machine learning (ML) is the scientific study of algorithms and statistical models that data scientists use to perform a specific task (without using explicit instructions that rely on underlying, often latent patterns) and to make automated inference over a period of time (Bishop 2006).

While there are many classifications of machine learning, in this book we classify machine learning into three broad types: unsupervised learning, supervised learning, and semi-supervised learning. This classification is not exhaustive, and, in fact, there are many other, more complicated ways to think about machine learning systems, such as reinforcement learning and association rule learning. However, for the purposes of this introductory text, our three-part classification will suffice. As we learned in chapters 2 and 3, one major issue that arises in the data science pipeline is where the data comes from, how it is represented, and what kinds of biases exist within it (or were added to it during data preparation). We also learned that in the context of machine learning, data is often "labeled" or "tagged" with "ground truth"—for example, labeling as a picture of a cat versus a picture of a dog, or labeling as a "virus-symptom cough" instead of a "bacteria-symptom cough."

Thus, we use the extent to which this labeling exists within datasets as a useful and accessible way to categorize machine learning models.

Unsupervised Machine Learning

Unsupervised machine learning looks for previously undetected patterns in a dataset with no preexisting ground truth labels and with minimal human supervision (Kassambara 2017). Often, this is the cheapest way to collect data, as labeling often costs money. There are many different modeling approaches within unsupervised machine learning. We want to outline a few popular approaches, recognizing that there are many others that we do not represent here. Chapter 6 shows how these models can be enhanced with human context and methodological approaches from social science and design.

- *Cluster analysis* groups a set of objects such that objects in the same cluster are more similar (in some way) to each other than to objects in another cluster. There are many approaches to clustering (e.g., density-based, centroid-based, distribution-based, hierarchical) depending on how you want to ascertain similarity. A widely used, simple example of centroid-based clustering is *k-means*, where we want to partition n observations into k clusters in which each observation belongs to the cluster with the nearest mean. Figure 4.1 illustrates k-means clustering. K-means is useful because it

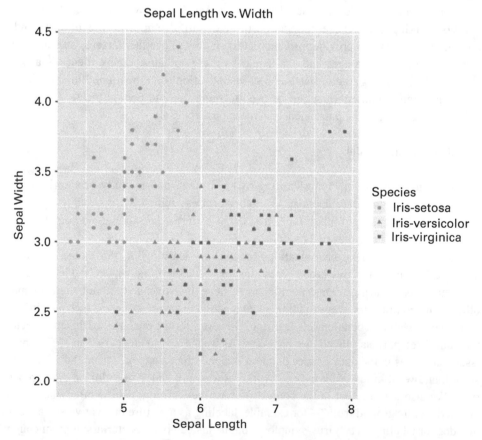

Figure 4.1
K-means clustering of Fisher's *Iris* dataset.

is simple to calculate, is easy to understand, and has multiple applications in science and industry. An example is clustering subspecies of the same flower.

•   *Principal component analysis (PCA)* is a way to reduce the complexity of datasets from many possibly correlated variables to a few uncorrelated variables by doing some data transformations (usually orthogonal). The hope is that these uncorrelated variables represent underlying latent variables that embody "true" theoretical measurement. These latent variables can be related to the observed data both linearly (common and easy to interpret) or nonlinearly (uncommon and harder to interpret). This is a form of dimension reduction, or a way to reduce the complexity of data to manageable, interpretable forms. Figure 4.2 illustrates PCA. PCA is useful because, compared to other ways of reducing complexity in datasets, it is easy to compute and visualize. Hence, it is very popular in applied data science fields like finance. However, PCA also has several limitations; notably, there is an underlying mathematical assumption that the latent variables or components are not correlated to each other, whereas in real life they often are, which presents issues with interpretation and validity.

•   *Topic modeling* categorizes large text corpora into overarching "topics" or distributions of words that occur with statistical regularity. The hope is that these statistically regular "topics" represent underlying semantic structures in the text or, more colloquially, broad topics or themes that arise from this text. Figure 4.3 illustrates one of the most common topic modeling approaches, *Latent Dirichlet Allocation (LDA)*. While there are many other topic models, LDA is widely used because it works well for large text corpora and hence is very popular among humanities and social scientists. On the other hand, it can be hard to interpret, as outcomes are provided as

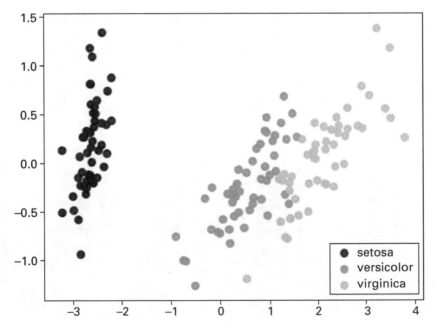

Figure 4.2
Principal component analysis of Fisher's *Iris* dataset.

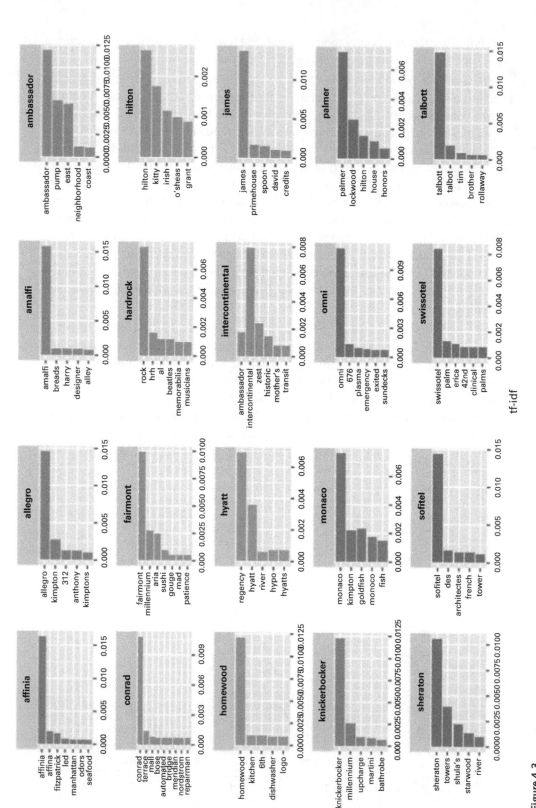

Figure 4.3
LDA topic modeling output example.

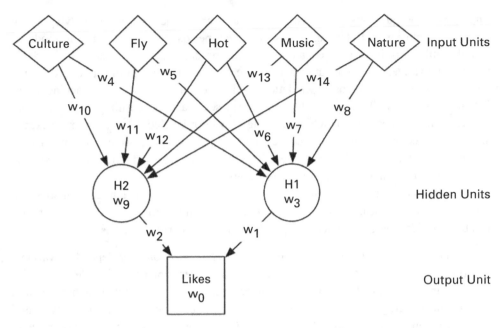

Figure 4.4
Artificial neural network for vacation recommendations.

strings of text without context. Additionally, LDA is not very effective for short pieces of text, such as a large dataset of 140- or 280-character tweets.

- *Artificial neural networks (ANNs)* or, more simply, *neural networks* consist of a network of "neurons" and the edges that connect them. If a given neuron receives a signal, it processes it and alerts its connecting nodes to also process the signal. In this way, a network of neurons can pass complex information through several intermediate layers before producing a final output or result. Neural networks comprise a large family of machine learning models that are used very frequently in data science. One unsupervised example is the *self-organizing map (SOM)*, which can take complex, multidimensional data inputs and create a two-dimensional map that effectively acts as a dimension reduction tool very similar to the output of PCA described above. Figure 4.4 illustrates how a SOM works. SOMs can reduce complexity and reveal underlying latent variables that might be good measurements of theoretical constructs.

Supervised Machine Learning

Supervised machine learning maps an input to an output based on input-output pairs (Maglogiannis 2007), where the input is a vector (a series of values) of *predictors* (features) and the output is usually a single value. Essentially, this refers to data that is labeled or tagged with some notion of "ground truth" (see chapter 3), usually by a human being. As we discussed in chapter 3, there is wide disagreement about what constitutes "ground truth" and how it is defined in different settings. For example, clinicians (i.e., domain experts) might decide what constitutes ground truth for data science models in healthcare settings. By contrast, in studies of Twitter data, crowdworkers or undergraduates may be asked to label tweets. How disagreements among labelers are handled can have a significant impact on subsequent results (Chen et al. 2018). Labeled datasets are usually split into "training" datasets and "testing" datasets. Any supervised machine learning model is usually built on training datasets and run on testing datasets to make sure the model is reliable and valid. Additional data that doesn't come from the same time or data collection medium is often collected as a "validation" dataset to make sure that the machine learning model works well on a new dataset. Supervised machine learning is usually more accurate than unsupervised machine learning. However, labeled datasets are more expensive to produce and often hard to collect. This presents an interesting trade-off that we discuss in chapters 5 and 6. We briefly describe some commonly used supervised approaches here, but keep in mind that there are many others.

Many supervised machine learning models are borrowed from standard statistics and illustrate how complicated and blurry the lines between machine learning and statistics can be. Consider these examples of supervised machine learning models:

- *Linear regression* maps the relationship between a response or dependent variable (ground truth) and one or more (usually more) predictor or independent variables (features). Statistically, the objective is to produce the best-fit straight line that produces the most average outcome. Linear regressions can be used for both classification and prediction tasks. As with all models, there are many variations to linear regression. The most common machine learning application of linear regression is to add a variable or function that provides additional information to reduce bias, especially if we are working with large amounts of data. Linear regression is used widely in many data science application areas, because it is simple and intuitive to understand and interpret without needing a lot of background mathematical sophistication. Figure 4.5 illustrates linear regression.

- *Logistic regression* models calculate the probability of a particular event occurring (e.g., pass/fail) given a set of predictor or independent variables. These models are

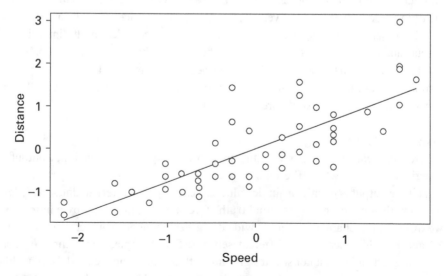

Figure 4.5
Linear regression on cars dataset (speed versus distance).

usually used for binary prediction tasks, although models that predict multiclasses as well as ordinal classification (i.e., ordered categories such as "low-medium-high") also exist and are widely used but harder to interpret. As with linear regression, logistic regression models (especially binary logistic regression models) are widely used because of their relative simplicity. Figure 4.6 illustrates logistic regression.

- *Decision trees* attempt to predict a specific value (or range of values) for a given dependent variable based on a set of predictor or independent variables. Broadly speaking, a decision tree consists of nodes in a hierarchy. Usually, each node represents some decision point (e.g., age > 5). Figure 4.7 shows the structure of a decision tree with all its constituent parts labeled. This model builds an optimum decision tree where each nonleaf node is labeled with an input feature (e.g., age < 50). The arcs coming from a node labeled with an input feature are labeled with each of the possible values of the target or output feature, or the arc leads to a subordinate decision node on a different input feature. Each leaf is labeled with a class or a probability distribution over the classes, signifying that the dataset has been classified by the tree into either a specific class or a particular probability distribution (which, if the decision tree is well constructed, is weighted toward certain subsets of classes). Decision trees are used in both classification and prediction tasks. They are commonly used because it is easy

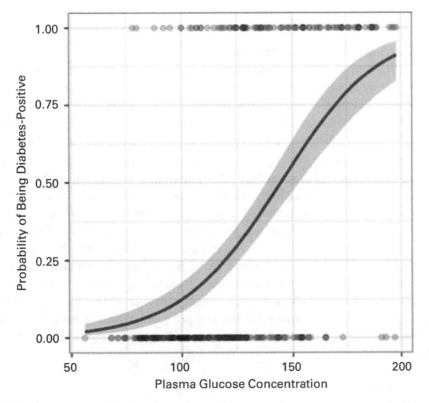

Figure 4.6
Logistic regression on a diabetes dataset.

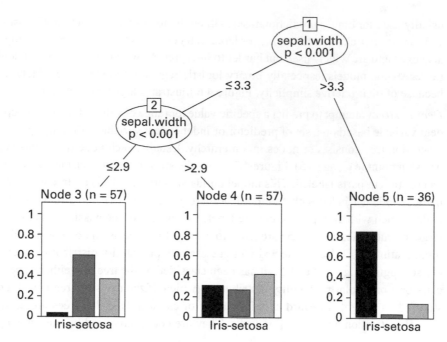

Figure 4.7
Decision tree example from Fisher's *Iris* dataset.

to visualize decision points (and, hence, the predictive power of a variable) within nodes. On the flip side, decision trees can become significantly unbalanced without more complicated corrections. One easy fix is to run a decision tree model many times and pick out the most common decisions. This is known as a *random forest* and is usually almost always a better analytical approach than a singular decision tree.

• *Support vector machines (SVMs)* aim to construct an optimum straight line (strictly speaking, a hyperplane in higher dimensions) as a "boundary" that separates sets of points that are labeled as belonging to different categories. When new labeled or unlabeled data points are added to this model, it automatically classifies them as belonging to one category or the other based on which side of the aforementioned "boundary" they are a part of. SVMs are primarily used for binary classification problems, although there are more complicated extensions that can do prediction or multiclass classification tasks as well. SVMs are often used for binary classification with lots of data points because they perform very well and can be visualized well. However, their inner algorithmic mathematics can be quite complicated to understand. This makes SVMs a great exemplar of machine learning models that provide excellent outputs without much transparency or explanation as to how they actually work. See figure 4.8 for a visualization of SVMs.

Semi-Supervised Machine Learning

Semi-supervised machine learning sits somewhere in between unsupervised and supervised learning problems (Chapelle, Schölkopf, and Zien 2010; Zhu and Goldberg 2009). Given the inherent trade-off between cost and performance, semi-supervised machine

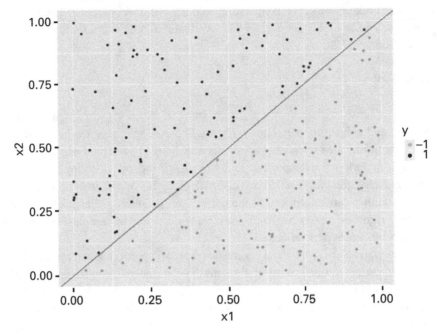

Figure 4.8
Support vector machine example on random data.

learning tries to balance building models when only some but not all data points are labeled. In practice, semi-supervised machine learning often deals with problems where only a small amount of labeled data is available. We outline a few approaches with the goal of introducing you to some of the most widely used applications. A caveat here is that semi-supervised models are generally not covered in introductory data mining or machine learning courses, as they are more mathematically sophisticated than either unsupervised or supervised machine learning models. We discuss two main approaches to semi-supervised learning:

- *Generative models* can be viewed as an extension of supervised learning (classification plus information about some probabilities of belonging to a given class) or as an extension of unsupervised learning (clustering plus some labels). In contemporary generative work, the algorithm "learns" the patterns in a dataset and can then create novel instances (records) that have the same pattern as the learned records. For example, there is an algorithm called Bach Doodle that can learn the patterns in Johann Sebastian Bach's musical compositions (Huang et al. 2019). A user can provide a brief theme (a tune, a sequence of notes), and Bach Doodle will then produce a novel musical composition around that theme, in Bach's style.

- *Graph-based models* develop a graph representation of data where each node is either a labeled or unlabeled example. Graph similarity methods like centrality metrics or other distance metrics are then used with the intuition that "closer" labeled and unlabeled pairs probably belong to the same category and hence can be classified into the same class. Shion and Michael used this kind of representation to show that IBM employees with more social connections to other employees (higher centrality) were more likely to be highly engaged in their work and in their attitudes toward the workplace (Guha et al. 2016).

Statistical Models

In the broadest sense, statistics is the scientific study of collecting, organizing, analyzing, interpreting, and presenting data (Lowry 2014). Some people have claimed that statisticians are the original data scientists (before machine learning as a field was even invented, even though the latter is more popular today in imagining and thinking about data science). Regardless, statistical models are also fundamental to and overlapping with data science. While there are exceptions to the norm, the discipline of statistics is concerned with making rigorous inferences from a sample (or a series of samples) that represents some underlying larger population.

In general, statisticians think about parameters that can appropriately represent data (e.g., *parametric statistical models*). Some of the common parameters are mean and variance, but many esoteric and mathematically sophisticated parameters can also be considered. However, parameters that describe underlying data distributions in a sample can artificially constrain data in different ways and might not allow us to make inferences from a sample to a population. Thus, *nonparametric statistical models* can be useful because they make fewer assumptions about the parameters of the data distributions (e.g., mean, variance, symmetry). Finally, to collect samples in the appropriate ways to make rigorous inferences about a larger population, we need to think about designing experiments.

All of this sets up what is known as *statistical hypothesis testing*, which is the formal name for the inference-making process. Statistical hypothesis testing often takes many forms, such as testing to see if a sample is actually representative of an underlying distribution, or if two or more samples are similar to an underlying distribution, or if two or more samples are similar to each other.

The two most popular families of parametric statistical models are the frequentist and Bayesian statistical models, which also have nonparametric variants. We want to introduce you to these models, then provide a brief introduction to experimental design and the need for appropriate experimental design to do good human-centered data science with standard statistical models.

Parametric or Frequentist Statistics
Parametric statistical models (also called *frequentist statistical models*) are possibly the most widely used family of parametric statistical models (Lowry 2014). The general idea is that there are hypothetically "true" parameter values (e.g., mean and variance) for a given population. That is, data (samples) are not constant, but the true parameters "underneath the data" are constant. Each dataset is considered to be an imperfect example ("sample") from that underlying "true" data, and we use those imperfect samples to estimate imperfect values of the "true" parameters of the "true" data. Therefore, we keep drawing samples from a population and comparing them with these theoretically true values to find either (a) that a statistically significant difference exists between the drawn

Case Study 4.1
Patient-Reported Outcomes Measurement

Brooke Magnus, Boston College

Healthcare providers are very concerned with understanding if certain policies have positive or negative outcomes on the overall health of patients. Historically, this has often been measured with biomarkers (e.g., blood pressure and cholesterol levels for heart disease). However, researchers have shown that just depending on biomarkers can be inaccurate (Pakhomov et al. 2008). Therefore, many healthcare providers ask patients about their own perceptions, symptoms, and opinions as an alternative approach, especially for conditions such as depression or pain that are hard to directly measure with biomarkers. This strategy is known as patient-reported outcomes (PROs) assessment.

PROs can generate a lot of messy and complex data that is difficult to analyze using traditional approaches. Data scientists have turned to the discipline of psychometrics to help measure and analyze this complicated data. Psychometrics is the science that develops measurement "instruments" (e.g., surveys and questionnaires) and statistical models for complex, human data that is hard to directly measure or analyze. Psychometric methods as a part of data science can play a critical role in our understanding of these kinds of health outcomes.

The US government is very interested in understanding health outcomes of people and therefore, through the National Institutes of Health (NIH), has sponsored the multiyear, multisite Patient-Reported Outcomes Measurement Information System (PROMIS) project that I have worked on for the past few years. The objective of this project is to develop a standardized set of human-centered measurement instruments and statistical models for use across a variety of clinical research and practice settings. The primary statistical technique underlying PROMIS is known as item response theory (IRT). IRT assumes that a patient's responses to a survey or questionnaire are related through an unmeasured, hidden reason, and therefore an individual's response (and hence, their personal healthcare outcome) can be predicted based on this underlying reason. This hidden, underlying reason that is often very hard to measure directly is also known as a construct.

IRT places all PROMIS constructs on a common, shared scale (like a thermometer) that makes it possible to make group- and individual-level comparisons across different health domains and chronic diseases. These kinds of comparisons are very useful for understanding patient healthcare outcomes. PROMIS measures have been validated for many uses across diverse settings and populations, including with healthy and chronically ill adults and children. For example, validation studies have examined PROMIS measure responsiveness to change over time (Pilkonis et al. 2014), clinical utility across a range of chronic diseases and conditions (Cook et al. 2016), and comparability across different modes of administration (Magnus et al. 2016).

Qualitative methods are also key in the development and evaluation of PROMIS measures. Focus groups and item review panels comprising subject-matter experts and people from diverse backgrounds examine the content and appropriateness of surveys and questionnaires, paying particular attention to potential cross-cultural conceptual or linguistic challenges or biases.

Finally, PROMIS has resulted in the development of an algorithm that tailors a questionnaire for an individual patient and their particular health conditions, so a person's score can be tallied in under a minute after they answer only a few questions. This has clear implications for personalized medicine. For example, clinicians can instantly view patient responses, compare them with a reference population, and engage in shared decision making about treatment plans with the patient, ultimately improving the quality of their care (Baumhauer 2017).

Given the enormous scientific efforts behind the development of its 300 measures, PROMIS is often used to assess patient-centered health outcomes. The initiative has facilitated

(continued)

Case Study 4.1 (continued)

comparisons between different groups of patients. PROMIS measures have also helped improve patient care, allowing for more personalized medical decision making. But when we think of big data in the healthcare industry, we often think of electronic health records, healthcare claims, data from wearable devices, and other types of patient-generated health data. Despite all the empirical work supporting the widespread use of PROMIS measures, we are just starting to see AI and machine learning techniques applied to PROs. Such techniques, in conjunction with existing psychometric methodologies, may be able to extract hidden information contained in PROs that is not reflected in other types of big data. For example, PROs can be integrated into data collection tools to detect changes in an individual's health status and predict an appropriate course of action (Brichetto et al. 2020). With the push toward patient-centered healthcare in recent years, PROs and psychometric methods could play a significant role in human-centered data science.

samples and populations or (b) that statistically significant differences exist between two or more samples. Many common frequentist statistical models are also used within the machine learning frameworks discussed previously. We have already described two major frequentist methods in the earlier section on supervised machine learning:

- *Linear regression*, a frequentist statistical model that assumes a Gaussian distribution in the underlying data, has found application within machine learning.

- *Logistic regression*, a frequentist statistical model that assumes a binomial distribution in the underlying population, is also widely applied within machine learning frameworks.

In addition, *analysis of variance (ANOVA)*, a family of statistical modeling approaches and their associated estimation procedures (such as the "variation" among and between groups), is often used to analyze the differences among group means in a sample. This is useful for nuanced exposition of differences between different group or subgroup members in a given population, which is a very common data science problem in many different application areas.

Bayesian Statistics

The general idea behind *Bayesian statistical models*, a popular family of parametric statistical models, is that—unlike the frequentist approaches—there are no hypothetically "true" parameter values, like mean or variance, for a given population. Instead, we assume that parameter values keep changing from dataset to dataset (Bolstad and Curran 2016). That is, parameters change because datasets change. Essentially, as more information comes in, your predictions and probabilities get "updated," presumably to become more accurate over a period of time. The notion of a "machine" learning from different datasets jibes nicely with the Bayesian ideal of updating information over time as new information comes in. (However, Bayesian statistical models can also be used outside of the machine learning context.)

We should note that most introductory statistics courses in undergraduate and graduate data science programs do not teach Bayesian statistical models but instead prefer to focus on frequentist models. Bayesian modeling can get mathematically complex. We outline the intuition behind some of the most common approaches below.

Bayesian networks represent a set of variables and their conditional dependencies using nodes and edges in a graph. They are ideal for taking an event that occurred and predicting the likelihood that any one of several possible known causes was the contributing factor. For example, a Bayesian network could represent the probabilistic relationships between diseases and symptoms. Given symptoms, the network can be used to compute the probabilities of the presence of various diseases. Bayesian networks have a wide range of applications. Figure 4.9 depicts an example of a Bayesian network.

Bayesian hierarchical modeling is a statistical model written in multiple levels (hierarchical form) that updates the probability of parameter values using the Bayesian method. The intuition behind this kind of modeling is that oftentimes, in dealing with complex data and data science problems, we are forced to recognize that entities exist in nested and overlapping groups. Thus, when new data or evidence comes in, probabilities should get updated, conditioned on these hierarchical group structures. Bayesian hierarchical modeling is frequently used in industry by data science research teams, although it can be complicated to imagine and mathematically estimate. Figure 4.10 illustrates Bayesian hierarchical modeling.

Nonparametric statistics are a family of statistical models that make fewer assumptions about parameters in the underlying data (Corder and Foreman 2014). This means

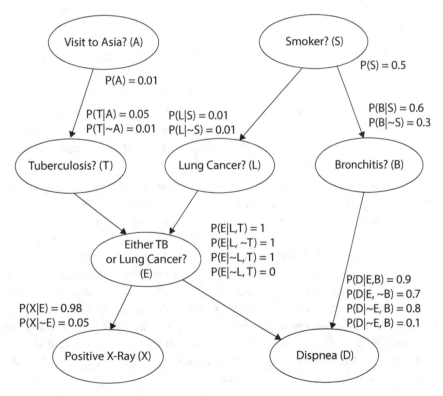

Figure 4.9
Bayesian network from a smoking cessation dataset.

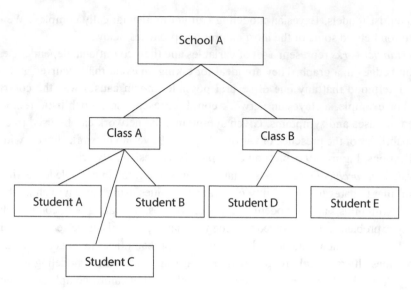

Figure 4.10
Hierarchical Bayesian model example.

that they are very useful for small amounts of data or data with outliers or skewed distributions. There are many ways to think about nonparametric statistics; here we present some that are often used in data science:

- Statistical tests that compare two different distributions: There are many examples, such as the *Anderson–Darling test* or the *Kolmogorov–Smirnov test*, which can be used to compare the characteristics of two disparate distributions.

- Statistical tests that compare two different samples relative to an underlying distribution: One of the most popular such tests in data science is the *Wilcoxon signed-rank test*, which can be used to compare the averages of two different samples drawn from the same underlying population.

- Nonparametric equivalents of parametric statistical models: For every popular and useful parametric model, there is a nonparametric equivalent. These can be useful when the underlying distribution is skewed or when there are good reasons not to use parametric tests. A great example is the *Kruskal–Wallis test*, the nonparametric equivalent of the ANOVA (discussed above), to test if two or more samples originate from the same distribution. This is very useful for comparing the means (or means of levels of categorical variables) to understand which ones are similar. The difference between ANOVA and Kruskal–Wallis is that ANOVA makes different, more restrictive assumptions about the underlying data. As a result, it is often more powerful than the nonparametric Kruskal–Wallis version of the same conceptual idea. However, for data that does not meet the statistical requirements of an ANOVA, the Kruskal–Wallis test may be more valid than an ANOVA.

Experimental Design

Experimental design is the process of formal analysis and design of experiments to conduct statistical modeling—be it frequentist, or Bayesian, or nonparametric in nature (Lindman 2012). In the context of data science, we can think about experimental design as an "anticipation" of supervised machine learning. A typical experimental design begins with a hypothesis that certain predictors (we would call them "features" in machine learning) can be used to predict an outcome (we would call the outcome the "ground truth" in machine learning). Historically, engineers and scientists who used experimental design were doing a form of supervised machine learning. They thought of experimental design not as a method to create a tool (i.e., a pipeline), but rather as a way to understand phenomena in domains such as biology, psychology, physics, and sociology.

Experimental design typically involves a series of tasks that aim to describe and explain the variation of information ("ground truth") under conditions that are hypothesized to reflect the variation. The term is generally associated with experiments in which the design introduces conditions ("features") that directly affect the variation, but it may also refer to the design of quasi-experiments in which natural conditions that influence the variation are selected for observation. In its simplest form, an experiment aims at predicting the outcome by introducing a change in the preconditions, which are represented by one or more independent variables, also referred to as "input variables" or "predictor variables" (we would call them "features"). The change in one or more independent variables is generally hypothesized to result in a change in one or more dependent variables, also referred to as "output variables" or "response variables" (we would call this class of variables "ground truth"). The experimental design may also identify control variables that must be held constant to prevent external factors from affecting the results.

Experimental design involves not only selecting suitable independent, dependent, and control variables, but also planning the delivery of the experiment under statistically optimal conditions given the constraints of available resources. There are multiple approaches for determining the set of design points (unique combinations of the settings of the independent variables) to be used in the experiment. We should note that experimental design courses are generally only available at the upper undergraduate or advanced graduate level in most data science programs. We don't cover experimental design in depth, but we have recommended readings at the end of this chapter. Many issues that arise because of a lack of good experimental design may lead to poor data science results, including false positives, insufficient statistical power, or p-hacking (Kay et al. 2016, 2017). Moreover, when dealing with human participants, issues of ethics, informed consent, confidentiality, and privacy must be addressed. We discuss these issues further in chapter 5.

Automated Models and Methods for Training Models

Until this point, we have described data science approaches that involve knowledge of the methods and careful examination of the data. Recently, several organizations have introduced tools that are intended to automate nearly all of the data science cycle (described in chapter 2), and that attempt to avoid the pitfalls described in chapter 3. These methods are variously called AutoML (Liang et al. 2019), AutoAI (Wang et al. 2020), or other project or product names (Kanter and Veeramachaneni 2015; Olson and Moore 2018). We refer to this group of methods and products by the more generic term AutoDS, for automated data science.

The goal of AutoDS is to allow a user who does *not* know data science methods to perform a simple, automated form of data science. In the simplest case, someone provides the data and the AutoDS tool completes most of the steps of the standard data science cycle, providing the "best" results according to built-in criteria, including a functioning pipeline that can receive new data and provide new outcomes from those data. Some AutoDS tools provide a ranked list of recommended solutions, along with evaluation criteria such as accuracy of predictions.

AutoDS tools are designed for people who are domain experts in a particular area but are not data science experts. Some data scientists are reluctant to use these automated tools because their features are not a good match for their needs for transparency and control (Wang et al. 2019). A follow-up study showed that the concerns of data scientists could be partially addressed through improved transparency of AutoDS operations, primarily through display of model performance metrics and visualizations of the statistical distributions of auto-cleaned data and a graphical summary of the automated operations along the data science cycle (Drozdal et al. 2020). Nevertheless, this field is still new and further research into the effectiveness of AutoDS methods is needed.

Overfitting and Underfitting

Two important and related concepts associated with fitting theoretical models to actual data are *overfitting* and *underfitting* (Handelman et al. 2019). As mentioned in chapter 3, data science models map inputs to outputs using various theoretical approaches that are outlined in the preceding sections of this chapter. Real-world datasets very rarely fit a theoretical model with a 100 percent accuracy rate. In fact, deviations from theory are expected and accounted for in almost all data science models. Overfitting occurs when a data science model (perhaps demonstrated by a curve) fits the observed dataset perfectly, such that the model represented by the curve passes through all of the observed data points. It might be counterintuitive since we always strive to attain the best fit in building models, but too much correspondence between a model and reality is also not desired. If a model is overfit, the implication is that the model only fits the current dataset and may be unable to fit additional datasets or predict future data reliably. This affects how generalizable our models are, and as we talk about later in chapter 5, generalizability is a desired trait in building data science models.

On the other hand, underfitting occurs when the chosen theoretical model does not appropriately represent the underlying complexities or distributions of the data at all. This is obviously undesirable because such a model will have very poor performance and predictive power, being unable to predict future observations reliably. In data science, we want neither overfitting nor underfitting but just the "best" possible fit under the circumstances.

Bias Detection and Mitigation Tools

As we have shown, it is possible for biases to occur in datasets and in the processing of data in a data science pipeline. This can happen even with the best of intentions. In this section, we describe some sources of bias, and we discuss ways to reduce, or "mitigate" some aspects of bias.

For example, different biases may exist in the discovery or capture of the data, the cleaning and wrangling of the data, the choice of predictors or features, the choice of the model, the outcomes, as well as biases that emerge unexpectedly.

Biases in data arise from multiple sources. For instance, bias can arise from choices made in data collection or sampling as well as decisions made to record data in certain ways, shapes, or forms. There are several points of concern here. First, before collecting data, people choose which data to collect and how. In data science practice, there is a tendency to gather data that is easily available and easily quantifiable. However, as with our discussion of the streetlight effect (in chapter 1), this may not be the best choice to answer the research question. How you decide to collect or record data may lead to irreversible biases. People can introduce biases into the system with their choice of sampling method, affecting representativeness. Finally, when you combine data sources (a very common practice), you might introduce another source of bias. Keeping records of these decisions helps with later analyses and with establishing the provenance of the data.

Biases in features can be introduced during data wrangling or cleaning and feature engineering. For instance, data wrangling and cleaning choices affect how features behave within the model. Feature engineering often excludes expert knowledge about the particular domain. In theory, the competencies, skills, and deeper understanding of the specific contexts of data science projects are supposed to be part of the feature engineering process but may be impractical to implement.

The computational process of feature engineering on large amounts of data, which is common in data science practice, is usually a form of statistical dimension reduction that results in computationally combined features being created that usually require a human domain expert to determine their meaning. What we hope for is that features end up representing some underlying concept or theoretical construct that can provide enough predictive power to develop a good model. Yet without actually knowing what this theoretical context is, there is no way to know if this mapping is actually useful or meaningful for analysis. This is especially true when features are constructed from lots of individual pieces of data that by themselves provide little predictive power. For example, a feature constructed from individual mouse clicks on a web page over a period of time might not contribute to a meaningful analysis.

Biases in models arise from a variety of sources. First, data scientists choose which modeling approach to adopt—for example, linear regression versus logistic regression. This is the primary source of biases in models. For instance, the chosen modeling approach may simply not be appropriate for the type of data or features that are available. Second, even when the modeling approach is appropriate, when constructing the model data scientists make choices that may introduce biases. For instance, data scientists choose which parameters of a model to change or keep—an "invisible" problem that can bias a model. Preventing problems that could arise from modeling and parameter choices requires not only computational and mathematical sophistication, but also an understanding of the context in which the features are being analyzed.

Biases in outcomes primarily arise from failing to ask critical questions around what outcomes are actually being predicted. For instance, are they the right outcomes to predict for the given question, or would other modeled outcomes be better for the questions at hand? For example, in child protection services, the emphasis is on predicting risks to children, while research in social work shows that attention to child–foster parent matching prediction has better outcomes (Saxena et al. 2020). Mitigating this requires deep contextual knowledge of which data science approaches are popular but also which data science approaches might need to be focused on to make progress.

Biases from other causes that have little to do with data, features, models, or outcomes can also emerge. For instance, bias could emerge if models developed for one context are applied in a different context. This is similar to the appropriation of data discussed earlier, but now we refer to the appropriation *of the model* or the pipeline. A great example is if crime prediction models are also used to predict child safety in welfare systems. On the surface, the prediction tasks look similar, but they actually represent very different problems with their own sets of contexts and issues that cannot just be applied in a blanket fashion. Or, for instance, circumstances and underlying data and their contexts might change over time but the model may remain stationary. This is sometimes called *model shift* or *model drift* and can cause heretofore-unknown biases to emerge.

Recently, there has been a focus on the development of bias mitigation and detection tools. All of these tools attempt to address bias from a computational and algorithmic perspective. Some of these tools are proprietary and internal to specific industries and are invisible to us beyond media reports, industry research papers, or internal white papers. Others—such as Aequitas (Saleiro et al. 2019) or IBM's AI Fairness 360 (Bellamy et al. 2019)—are freely available open-access tools that the broader data science community can use to mitigate bias. This is a dynamically changing field of applied research, and there are many questions as to the overall effectiveness of these products. Here are some broad guidelines surrounding the use of these tools.

- Remember that these tools are intended for users who are not data scientists. Someone working in data science would be well served to understand the mathematical and computational rationale for these tools. Many of these tools arise from recent academic work. These are active areas of research and practice, and it is important to find the latest version of the tool that you want to use and the latest discussions of that tool. Read the underlying literature that defines and explains the limitations of the tools before applying them.

- Understand the social science and context behind these tools. It is not clear that all of these tools are effective on all forms of bias—especially because the discussion of sources and mitigations of biases is ongoing, with newly recognized types of bias being continually added to our understanding. We should remember that all data science projects are different, with unique assumptions and contexts. A tool built for a particular context may not work effectively to remove bias in another context.

- When deploying a tool, it is best if you can quantify and visualize the bias that was detected or mitigated. Written statements and figures that explore alternative paths (if this bias mitigation is not done) should also accompany your data science project report to compare and contrast how these tools performed. For instance, you should carefully mention the types of biases and where within your pipeline bias was mitigated and what might happen in real life (with examples from your data) if you didn't do bias mitigation. It may be useful to compare and contrast data science projects with the same underlying data, models, and pipeline but differing in the application of bias mitigation and detection tools.

Human Decision-Making in Data Science Models

Human decisions affect how data science models are built. Here we provide a few examples of some of the technical decision-making that goes into developing a data science model:

- When doing unsupervised learning, data scientists often have to choose the initial objectives or goals of a model. For instance, in k-means clustering, you have to choose the exact number of clusters that the algorithm will find; in topic modeling, you have to initialize the number of topics that the algorithm will generate for you. This demands not only technical sophistication on your part but also a deep knowledge of the context, as we discuss in the next chapters.

- Data scientists must make decisions about changing technical parameters that make a model work. For example, in k-means clustering, you need to choose which "distance metric" to use to determine similarity of clusters. In regression modeling, you have to change internal parameters depending on the shape or distribution of data.

- Data scientists also have to decide on statistical ranges or cutoffs for many statistical models. Under many circumstances, these decisions are quite subjective. For instance, in doing principal component analysis, you need to decide the cutoff to determine how many "principal components" are included. Again, in topic modeling, you have to decide, based on a suite of metrics, which topics are of sufficient quality to keep in the final model and which should be excluded for being of "poor" quality. We will discuss some of the social and ethical implications of these decisions in chapters 5 and 6.

Visualization

A model is only as good as its representation and how well it can be communicated to the people who will use it. As we argued in chapter 2, visualization can be one of the most effective means to convey knowledge to others; in addition, it can be used as an important (and necessarily human-centered!) component of a data science pipeline.

Data visualization is the visual representation of data in the form of charts, graphs, and diagrams. Originally these visualizations were static and often created by hand with pen and ink. Now visualizations are computer-generated, dynamic, and interactive. With the rise of computational power, we still need skills to make good visualizations. As data visualization expert Stephen Few (2009) wrote, "Computers can't make sense of data; only people can."

Data visualization is often divided into three main categories: (1) *scientific visualization* displays data that has a physical position in space, such as a gas combustion dataset; (2) *information visualization* displays abstract information that may not have a physical reference, such as sales or stock prices; and (3) *visual analytics* is a combination of highly interactive visual interfaces and statistical learning algorithms. All three occur in data science. As we mentioned in chapter 2, the human visual system is the highest-bandwidth channel into the human brain, and thus visual representations of large amounts of data are often the most efficient and effective ways to convey the information present in very large amounts of data so that people can make sense of it.

Data visualization is the process of mapping data variables to visual attributes, also known as data encoding. It is an intrinsically human-centered process because it allows human understanding of complex and large datasets. To map data variables to visual attributes, it's important to understand a set of key design guidelines and techniques used for the visual display of information, including their relationship to human perception.

There are many reasons to create visualizations. We might need to answer questions, make decisions, see the data in context, find patterns, present an argument, tell a story, or inspire, for example. Very large amounts of data may need to be represented in an interactive way to make them understandable.

Data encoding means classifying each variable into a type and then determining which visual attributes best represent these data types. *Data types* can be nominal, ordinal, or quantitative. Nominal data consists of labels or names, such as the names of fruits: apples, oranges, bananas. Ordinal data possesses an intrinsic ordering, such as large, medium, and small. Quantitative data consists of numeric amounts, such as length, mass, area, volume, position, date, or time. These data types are then mapped to *visual attributes* such as position, size, color, shape, or orientation, among others. Decades of research have established that certain visual attributes are more effective at conveying quantitative data. People can most accurately detect differences in position, followed by length, angle and slope, area, volume, and, last, color and density. When assigning these attributes to your data, make sure to choose the attribute of *position* for your most important quantitative data. Save color and density for nominal data, for example.

This process of data encoding is more difficult than it seems. It is a challenge to pick the best visual encoding among all possibilities. One set of principles is described by Jock Mackinlay (1986):

- *Principle of consistency*: The properties of the visual attributes should match the properties of the data variables (i.e., one-dimensional data such as cost should not be represented by area or volume).

- *Principle of importance ordering*: Encode the most important information in the most effective way. Since humans can detect position most accurately, use it to encode your most important quantitative data.

- *Principle of expressiveness*: The visual attributes should express *all* the facts in the set of data, and *only* the facts in the data. For example, if you represent two quantitative data points using color, it is impossible to tell whether one data point is greater than the other. That means this visualization cannot express the facts.

- *Principle of effectiveness*: A visualization is more effective than another visualization if its information is more readily perceived than the information in the other. For example, if you are comparing auto prices with mileage, a scatterplot (figure 4.11) is more effective than a bubble plot (figure 4.12), where one of the variables is represented by the area of a circle.

Before you design a visualization, we encourage you to consider two things: First, who is the audience and what do they need? Second, what is the purpose of the visualization;

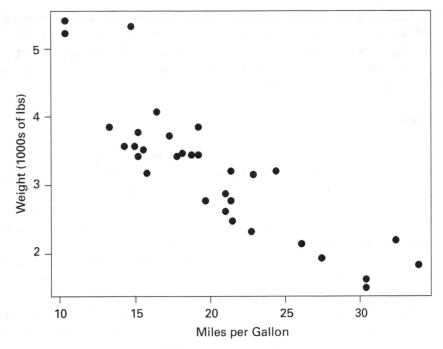

Figure 4.11
Scatterplot of cars dataset: weight versus mileage.

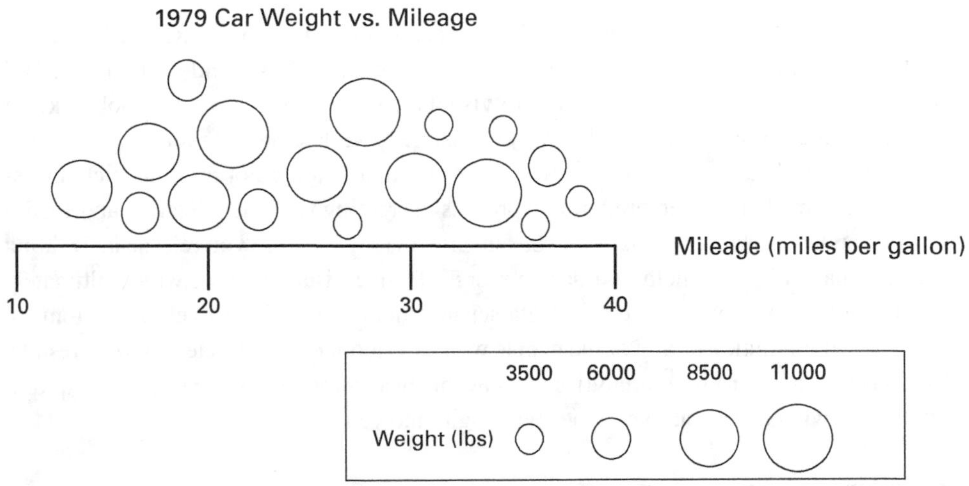

Figure 4.12
Bubble plot of cars dataset: weight versus mileage by horsepower.

that is, is it to explore or to communicate the data? Only after you have answered these questions should you select a visualization, such as a scatterplot, histogram, or bar chart. Also, consider your methods of interaction with the data. A full description of the data visualization process is beyond the scope of this book, but we have recommended several excellent texts on data visualization at the end of the chapter.

Visual Analytics

So far, we have primarily focused on visualization as a means of communication, but it is important to also consider the use of visualization as a powerful tool for understanding and exploring vast amounts of data. The field of *visual analytics*, first defined by James Thomas and Kristin Cook in *Illuminating the Path* (2005) as "analytical reasoning facilitated by interactive visual interfaces," has evolved rapidly over the last two decades as the sheer amount of data available in the world has grown exponentially, repeatedly overwhelming human comprehension. Interactive visual tools combined with machine learning and other analytical techniques have made it possible for humans to "synthesize information and derive insight from massive, dynamic, ambiguous, and often conflicting data." (Thomas and Cook 2005). Visual analytics tools have been used to debug complex data management software, to perform many vital human-in-the-loop evaluations within data science pipelines, and to "detect the expected and discover the unexpected" (Thomas and Cook 2005).

Visualization Tools

Visualization tools include software such as Excel, ggplot, matplotlib, R, D3, Power BI, and Tableau, among many others. They vary in power, ease of use, and cost, but they all provide some way to map data variables to visual attributes. Even though a tool makes it easier to create a visualization, it is always possible to violate one or more of the design principles described above. The best way to ensure that you are not making such a mistake is to follow a human-centered design process in creating your visualization and ensure that you conduct usability testing with your intended viewers before you release it. Training in data visualization will help you design more effective visualizations, which ultimately will enable you to communicate your data science models more effectively to customers, managers, communities, or any of the people who care about or are affected by your results. Using visualization or visual analytics tools as intrinsic components of your data exploration process will enable you to gain deeper insight into your data.

Conclusion

This chapter introduced several different types of machine learning and statistical models so that you can compare them for your needs. The main approaches to machine learning are unsupervised, semi-supervised, and supervised learning. Of the various approaches to statistical modeling, we have focused on parametric and nonparametric statistics, frequentist as well as Bayesian statistics, and introduced experimental design and a few methods common in data science projects. We also described emerging automated data science tools that may bring people without specialized knowledge or training into data science.

Bias mitigation and detection tools are important in data science, but there are limitations to using these types of tools. Most of the tools that are available to the general public are based on novel research that is often quite rigorous and mathematically sophisticated. They also might be challenging to apply without understanding the mathematics involved. Be cautiously optimistic about using these tools, as the existing scientific community hasn't quite coalesced around these tools; in fact, there is research suggesting that they are a lot more fragile than they appear.

Data visualization is one way to explore or communicate results of data science models, and we briefly mentioned the differences between scientific visualization, information visualization, and visual analytics. Visualization is a key part of the data science process, both as model representation and as a fundamental component of data understanding and analysis itself.

By foregrounding the people who do the work of data science, we see how decisions made during the data science modeling and representation process rely on people's mathematical and technical backgrounds as well as their interests, choices, biases, concerns and perceptions. You should endeavor to be reflective about these factors and their impact on your data science project.

The summary knowledge of the current state-of-the-art in data science tools and techniques provided in this chapter will give you a good background to understand the contributions of human-centered approaches to solving data science problems that we look at in the next chapters.

Recommended Reading

Chapelle, Olivier, Bernhard Schölkopf, and Alexander Zien. 2010. *Semi-Supervised Learning.* Cambridge, MA: MIT Press. This book is an excellent introduction to a complex and sophisticated topic. The common topics in semi-supervised learning are treated in a technical manner (assuming some introductory knowledge of machine learning) yet should be familiar and accessible to the reader who has taken a first course in machine learning and statistics.

Few, Stephen. 2009. *Now You See It: Simple Visualization Techniques for Quantitative Analysis.* Oakland, CA: Analytics Press. A practical and readable guide to creating effective data visualizations.

Lindman, Harold R. 2012. *Analysis of Variance in Experimental Design.* New York: Springer Science & Business Media. This book provides a good introduction to statistical analysis in traditional experimental designs.

Tufte, Edward R. 1983. *The Visual Display of Quantitative Information.* Cheshire, CT: Graphics Press, and Tufte, Edward R. 1990. *Envisioning Information.* Cheshire, CT: Graphics Press. Tufte is one of the founders of the modern incarnation of data visualization, and his books are beautifully produced works of art full of practical information on how to visualize data.

Ware, Colin. 2020. *Information Visualization: Perception for Design.* 4th ed. Cambridge, MA: Morgan Kauffman. This classic work by a pioneer in the field has a strong focus on the psychology and physiology of visual perception and how to leverage human strengths to make sense of large amounts of data.

Zhu, Xiaojin, and Andrew B. Goldberg. 2009. *Introduction to Semi-Supervised Learning.* San Rafael, CA: Morgan and Claypool. An introduction to modern semi-supervised machine learning practices. The book assumes readers will have some knowledge of introductory machine learning, statistics, and optimization, but provides a fairly exhaustive technical overview of the different and commonly used semi-supervised learning approaches.

5

Human-Centered Approaches to Data Science Problems

We have looked at the data science cycle (chapter 2), the different steps where human-centered concerns can and should be addressed (chapter 3), and the different types of data science tools and models (chapter 4). Now we zoom out to look at the data science process as a whole. In this chapter we aim to introduce you to what makes human-centered data science different at every step of the process from other approaches to data science. We show you four key practices:

(1) Anchor your project with good questions and start by first figuring out what you want to know rather than what data you have.

(2) Develop ethical practices as one of the most important aspects of human-centered data science.

(3) Think about how your projects may get used by others.

(4) Reflect on the data science process to improve your practices and to become better at data science.

The approach to data science that we show here concerns both the *process* of doing data science and the *context* in which one does data science. We emphasize that our ethical reasoning about our practice is inseparable from our practice. To paraphrase internet researcher Annette Markham (2006), ethics is method—method is ethics. Ethical data science is good data science, and good data science relies on practices that support ethical design, choices, and actions. The real power of addressing ethics in human-centered data science is that it shows how ethics is practice that is informed by principles.

Asking Good Questions

Good questions, not convenient datasets, provide the anchor for good data science projects. Having a question that guides a project helps you think about what data you need, how much data is enough, and how the data connects to the question you want to answer.

Different disciplines may emphasize this to different degrees. When Shion, whose academic background was in statistics, was a graduate student in information science, he began collaborating with a senior social scientist studying media and health. He wanted to predict social media nonuse in a large dataset. He could work within the categories of data to show statistical correlations. He crunched a lot of numbers and built what he thought at the time was an excellent, rigorous predictive model based on the dataset at hand. His faculty collaborator took one look at his initial results and asked him, "But what is your question here? What is it that you are trying to find an answer to?"

"I don't understand. How many people don't use social media?"

She replied, "That's a start. But how are you going to define and predict non-use [when people don't use social media]? . . . After all, there's no column label in the data that says non-use." The better way would have been to start with the question first, not the qualities and categories of the data. The end of the story was that this question led Shion and his collaborator to conduct a large survey to help understand how people perceive their own continuum of use and non-use. It also led them to several research publications on the topic. Getting more specific on the right questions to ask can lead to better and richer results.

This story is about learning how to *operationalize* what we are measuring in our projects and how we do it. Knowing that two categories are related, or "exploring" a dataset to see what relationships emerge, may be one place to start. Take a step back from the dataset to figure out what it is you want to ask and what it is you want to answer. Sometimes building toward answers requires other data, different data, or looking at the data in a different way.

Correlations are easy enough when you know how to do them. Good questions take work. Without a good question, those correlations cannot explain relationships or dynamics in human behavior. With good questions, our results leap from describing the relationships in our datasets to telling the story about the relationships and dynamics in the world—and explaining human behavior.

Good questions are meaningful. At the core, asking good questions is about figuring out what you want to know. This seems intuitive, and yet we find from our experience of teaching data science that asking good questions is one of the hardest things to learn how to do. Good questions ask something that we can find the answer to. Good questions ask something novel, something we do not yet know. Good questions are clear enough to be understood, specific enough to be doable with a data science project, and yet broad enough to capture something meaningful. Good questions are:

- *Empirical*: They ask something about the world.
- *Falsifiable*: We can answer them with evidence.
- *Focused*: We can answer them within our project.
- *Important*: We will be able to answer something significant with them.

We know interesting questions when we hear them: Do people in different countries share misinformation on social media at different rates? Does the structure of databases on climate information support more scientific collaboration on some datasets than others? Which matters more for helping sick people get out of intensive care—close collaboration among nurses or close collaboration between nurses and specialists? These are questions that Gina's colleagues and students recently asked and answered. Asking good questions helps us design better data science projects.

The most important quality for good questions is that they are important enough for someone else to want to listen to their answers. That means that the questions are "interesting." In 1971, sociologist Murray Davis shared a simple observation about interesting questions: they often follow a format that begins, "What seems to be X in reality is really non-X" (Davis 1971, 313). A more nuanced form of the question might be, "What seems to be X in reality is really non-X, but only for people with Y." We know women often face discrimination, but what happens when coders are anonymous on GitHub: Do women still face discrimination? (Vedres and Vasarhelyi 2019). We know Chinese social media faces governmental controls: Are there differences in what and how social media posts are removed (King, Pan, and Roberts 2013)? We think we know the answer, but with data and insights good questions may show us something novel or different.

Coming up with these kinds of counterintuitive questions is not straightforward. It requires knowing what people expect or assume. Counterintuitive questions "work" because they show how data science adds value—by helping a company better understand their customers, a newspaper to understand their readers, or a hospital to understand their patients. Without showing something novel, important, or interesting, it is difficult to justify the time and resources of a data science project. Good questions depend on knowing who thinks a question is a good one. It may be important for a newspaper to know that more people read the sports section on Tuesdays than Wednesdays, but that will not help a data scientist explain the question of *who* reads sports and why. Good questions help us think about the *why* of a problem as much as the *what* of a problem.

This practice of asking questions first is different from starting with a dataset and asking, "What might it be able to tell me?" Good questions guide data scientists through a process of defining and scoping their projects. Focusing on questions first is connecting to concepts and discoveries that are known and designing ways to find out the "why" of relationships. In psychology and other statistically driven social sciences, there is now a practice of "preregistration" of hypotheses to prevent what is called "p-hacking," chasing the small effects that inevitably appear in noisy data (Nuzzo 2014). It is easy to find something statistically significant in a large enough dataset, but we should be looking for effects that are *meaningful* within the context of everything else we know. And that usually relates to focusing on good questions first.

Asking questions first also helps us think through the ethics of formulating our project. It helps us in exploring what we *should* look for. Questions help us think about the problem or puzzle to explore in the project. This practice keeps us from only looking under the streetlight of where it is easiest or most convenient to look.

Ethics

Work in data science has the potential to harm people. *Ethics* refers to a set of principles that guide the behavior of a group or an individual to try to avoid harm and do good. In the United States, research on people—the "human subjects" of research—is governed by regulations that emerged from reports about unethical and harmful medical studies. *The Belmont Report* (National Commission for the Protection of Human Subjects of Biomedical and Behavioral Research 1979) holds three general principles that must guide research

Case Study 5.1
Ethical Ethnography in Data Ethics
Katie Shilton, University of Maryland

I did not set out to work in data ethics. I started grad school to be an archivist and work with paper. But I got rerouted thanks to research happening in my program, and a bit of luck: finding the UCLA Center for Embedded Networked Sensing (CENS). CENS leaders were some of the first researchers to use mobile phones to collect data about people. This was before the era of smartphones—at the beginning, we were working with Nokia feature phones with an external GPS device stuck to the back and a huge external battery pack needed to power the thing through a full day.

CENS became the setting for my dissertation research. I conducted a participant-observation study of how CENS engineers grappled with the ethics of the data collections they were performing. Importantly, I was learning to be an ethnographer alongside the engineers who were learning how to collect pervasive data (personal information generated through digital interaction, like social media data, search histories, geolocation data, and wearable device data). I remember telling a CENS leader that I wanted to observe CENS researchers for my dissertation, and he replied, "Like monkeys?"

People were both welcoming of, and weirded out by, my presence as an observer, and I had to do a *lot* to mitigate that weirdness and build trust. I regularly presented on my work and engaged CENS members in discussions of what I thought I was seeing. I did formal interviews that allowed participants to tell me their version of the story. I checked quotes with participants before using them in publications. I sometimes chose not to publish vignettes that I thought might be identifiable, or even potentially hurtful. These are lessons I took with me: that people often don't like the idea of being studied, and that it takes work to build the trust to do so. And that building that trust means both *being* trustworthy and demonstrating that trustworthiness to participants.

And as studying social media, smart cities and IoT [internet of things] devices became increasingly common in research, my dual roles merged. First, I was lucky to have been thinking about the ethics of this kind of data collection early. And second, I was lucky to be thinking about it as an ethnographer—someone who had to negotiate trust to collect lots of observational data. I was just collecting it differently (with my eyes and my pen, rather than with sensors or online traces).

Over the years, I found academic friends and colleagues who were also interested in big data and privacy, ethics, and risk. Together, we started the PERVADE (Pervasive Data Ethics for Computational Research) project to study empirical questions in data ethics. PERVADE began in the context of a growing public backlash against data science research. People were angry about a set of big data projects focused on people and their habits, such as the Facebook emotional contagion study, releases of analyses by dating sites such as OkCupid, and notoriously, the selling of research data to Cambridge Analytica to attempt to influence elections. This backlash didn't surprise me; it felt like a throwback to my ethnographic work. People often object to being studied without their knowledge.

Ethnographers have spent more than three decades questioning the ethics of their practice, the meaning of their data, and the position of their methods within larger movements (such as the ethnographers who participated in first European and then American colonization of other areas of the world). Data science now needs a similar reckoning, and I think that data scientists can learn a lot from ethnographers, those original embedded observers. Like ethnographers, data scientists need to reflect on at least two principles: awareness and representation. How aware are the people we are studying of data creation, and of our ability to mine it? Studying intentionally created Tweets is different from studying the traces left by our phone GPS. We need to consider mechanisms for public awareness (ranging from traditional forms of giving "informed consent" to public scholarship to participatory methods)

Case Study 5.1 (continued)

for particularly invisible or unobtrusive data collection. Second, data scientists should explicitly grapple with representation (i.e., who is included in the data, and who decides how their data is measured, structured, and recorded), and also power. We need to recognize when our data have historic biases or re-create injustices. We should use data science to increase the representation of those who have traditionally been left out of (or actively harmed by) research, and to hold our public institutions and powerful actors accountable. And we should reflect on the ways data science is embedded within larger social structures such as surveillance capitalism and historical inequities. We plan to use the PERVADE project to further each of those conversations.

with people, and these principles shape how research ethics are talked about today in university and medical settings: respect for persons, beneficence, and justice.

- *Respect for persons* means that research should respect people's rights to exercise autonomy.
- *Beneficence* means research should not harm people and should, if possible, benefit them.
- *Justice* means that the benefits of research should be distributed fairly.

Society expects researchers to uphold these principles, and academic research is required to meet these standards. These principles are found in research ethics guidelines and regulations in many countries, even if the specific local regulations and laws differ. While there are different rules in different countries and cultures, we can build on these principles and uphold them in a way that expresses ethical pluralism—the notion that ethical concerns may vary around the world—and cross-cultural awareness of these differences.

Within these cultural differences there are different emphases on what constitutes *ethical*. In some countries, concern for individual autonomy guides how research should be conducted. In others, like the United States, some tolerance for "reasonable" risks to autonomy is granted for projects that might otherwise be expected to benefit the greater good (franzke et al. 2020). The Association of Internet Researchers created a document for conducting research online with people's social media and other digital data and for doing ethical research with artificial intelligence/machine learning (AI/ML) (franzke et al. 2020), and it guides our thinking on ethical practice in data science. The ethical guidelines they produced provide a way to think about projects, taking into consideration a wide range of potential scenarios and contexts rather than an overly simplified checklist or a rule-based approach for ethics, especially as it fits to one place.

Ethical action is guided by thinking through *how* research is done and what *context* it is done in. Human-centered data science relies on this approach to ethics as a practice: ethics is based on not just what people formally are supposed to do or the definitions of those actions, but what people do in practice. Ethics is not a checklist, because it is impossible to devise a set of rules to fit all situations. Ethics is not about a goal for people to optimize or to "game." In our view, ethics means being reflexive at every step—thinking of what could go wrong, who could be harmed, and who could benefit. Ethics is not just reflection, though, and simply being reflexive will not necessarily prevent harm. You may need to consult with others—especially with people whose data is being used in your analysis. As we show later in this chapter, each stage of the data science cycle can raise new questions and challenges for ethical concern.

There are different approaches to ethics. *Care ethics* involves thinking through ethics as care for others, not simply as trying to minimize harm to others or to maximize benefit. Cultivating the ability to put ourselves in the position of others helps us to design data science projects as if the people involved were our families and loved ones. Would that change our approach to the decisions that we make?

An Example of Ethics Practices and Principles to Develop and Extend

With these considerations in mind, we present one of many different sets of ethical guidelines for people in data science to develop and extend. These guidelines follow the Association of Internet Researchers 2020 Ethics 3.0 guidelines and principles We encourage our readers to read the whole document (franzke et al. 2020).

Respecting autonomy means that data science projects should respect people's rights to privacy and to making decisions that impact their lives, because datasets provide proxies for human subjects and their behavioral patterns.

Informed consent means working to ensure that people have the power to choose whether to participate in your data science project, as either the intended or unintended subjects of the research.

Legal frameworks and regulations govern laws about people's privacy. People's rights to data about them in large datasets varies by country (e.g., the European Union's General Data Protection Regulation, or GDPR, for privacy protection), by context (e.g., the US regulation on healthcare data and privacy, known as HIPAA), and within the United States by state (e.g., the California Consumer Privacy Act or CCPA). Doing ethical data science means being aware of and following the applicable laws. However, reflection is also necessary because what is minimally "legal" may not always be ethical.

Third-party data is important for many data science projects. When data scientists use existing datasets that were collected by parties other than the data scientists themselves or the people represented by the data, it is important to consider the relationship of power and responsibility with regards to how this third-party data was collected and what responsibilities data scientists have to the people represented in the data we work with or who are affected by the decisions that will be made with our results. We must also consider the potential impact of third-party data processing and the ethical risks that potentially poses. Europe's GDPR rules are particularly strong and specific about who processes the data.

Freely available data is not free. The rise of enormous amounts of data gathered about people in their everyday life exposes us to potential harms of reidentification and of products and services being designed that may present different levels of benefits or potential harm.

Data governance means how we take care of the data we are entrusted with and who governs the rights and responsibilities for handling data. Several well-publicized cases, such as the Cambridge Analytica example above or Latanya Sweeney's research on Massachusetts health care data we discussed in chapter 3, show how easy it is to expose sensitive private information through data science projects.

Case Study 5.2
Data Is People: The Ethics of Scraping Data

Casey Fiesler, University of Colorado

When you ask people whether it is ethical to scrape data from a website, there are two common responses: (1) Does scraping data violate Terms of Service? and (2) Is the data "public"?

Whether it is unethical for researchers to violate Terms of Service (TOS)—particularly data collection/scraping provisions—has been a topic of debate for years. Both the law and ethics on this topic are fuzzy. However, the problem with a TOS-based ethical decision on collecting data is that it assumes that (a) violating TOS is inherently unethical and (b) violating TOS is the *only* thing that could make collecting data unethical. I suggest that neither of these is true.

There may be situations in which violating TOS against the wishes of a company could be an ethical act. For example, in a recent court case, researchers claimed that TOS violations are necessary for algorithmic audits that can help uncover discriminatory practices—and the court agreed with the researchers. Another concern is that if research on a platform can only be conducted by researchers explicitly given access to that platform, this might distort scientific discovery, particularly if researchers may be constrained by the company that employs them. There are also many sites that *don't* prohibit data collection, but where the research might have ethical problems for any number of reasons. Even well-intentioned research or application might be harmful, regardless of legality.

In a recent research project, my collaborators and I analyzed the data scraping provisions from over 100 social networking sites and found that these provisions are vague, highly inconsistent across sites, and most importantly, almost entirely lacking in context. With the exception of a few vague mentions of "personal" data, most data collection provisions are entirely context-agnostic when it comes to, for example, what the specific data being collected is, who is collecting it, what it is being used for, and what the expectations and potential harms might be for the people who created that data.

However, if you are making decisions about collecting and using data that was created by people, context is critical. This brings us to the second common answer to the question, "Is it ethical to scrape data?" Unfortunately, for many people, the only context that matters is: "Is it *public*?"

"Public" is the magic word when it comes to research ethics. "But the data is already public" was the response from Harvard researchers in 2008, when they released a dataset of college students' Facebook profiles, and from Danish researchers in 2016, when they released a dataset scraped from OkCupid.

However, as with data scraping provisions, the "publicness" of data does not tell us what really matters. The idea that a tweet about what someone had for breakfast is exactly the same as a tweet revealing someone's sensitive health condition is absurd. After all, one of the potential harms of using public data is *amplification*—that is, spreading content beyond its intended audience. It probably will not do much harm if more people know about your breakfast cereal. Negative consequences might follow a tweet about someone having cancer, a sexually transmitted disease, or depression.

What you plan to *do* with the data also matters. Consider the example of Cambridge Analytica, where data collected for research was used for more than subjects were told—not for science, but for manipulating elections. Data scraped from public social media sites and dating profiles has also been used to create predictive machine learning algorithms designed to classify sexual orientation (to use the language of the project) or inappropriately label gender.

As researchers, we have a responsibility to acknowledge that factors like the type of data, the creator of that data, and our intended use for the data are important when it comes to collecting and using it. However, these factors are harder to judge and understand than "Does this violate TOS?" or "Is this public?"

It can be frustrating not to have explicit rules to follow. However, ethics is not necessarily so much about following rules as it is about being thoughtful and carefully examining each situation contextually. This requires care and often extra work, but it is important to remember that data is often not just *data*; it also represents people.

Expectations mean that people may have different expectations about how their data is used that may not be reflected in the terms and conditions of their agreements or regulated in the law. For example, when people post on social media, they may expect that their privacy is otherwise respected, even though they may have consented to particular uses of that data when they registered for their social media accounts. A survey of Twitter users found that the majority of respondents felt that "researchers should not be able to use tweets without consent" (Fiesler and Proferes 2018).

Research aims and risks of harm mean that the principles of respect, beneficence, and justice can be applied when we have clearly scoped data science projects with good questions. Clarifying the aims of the research means that they can be weighed against the potential harms.

Data science is a responsibility, to us and others, the communities we study, and society at large. People working in data science should take the time to be thoughtful about this responsibility.

Ethics is about practices guided by principles. Doing good human-centered data science is about thinking through the specific challenges of any project, not simply having a standard format for every project. There is no recipe or checklist that can solve every ethical dilemma.

Ethics means asking the right questions. Without a clearly scoped project and focused questions, data science projects can generate unnecessary risks to people through potential reidentification, exposure to risks of loss of privacy, or harm through classifying people into categories that have real-life consequences.

These ethical guidelines were based on careful thought and consideration, and there are other ethical guidelines from different perspectives, including from Indigenous peoples' or First Nations' points of view. These guidelines provide another way of *framing* the situation of racial and cultural minorities in their own languages and in their own ethical contexts (Smith 2013). The act of framing is important because a conceptual frame (the way you think about a person or a group) can powerfully influence your decisions about how to include or exclude a particular group's data and how you combine or distinguish their data, in relation to other groups. The Government of Nunavut published a set of principles for education that includes an extensive discussion of Nunavut normative ethics, organized into six major legal areas and detailed ethical principles within each area (Nunavut Department of Education 2007). Suvradip Maitra (2020) lists five Indigenous principles for the governance of AI that depend on relationships and relational ways of knowing. Reporting for the Canadian Commission for UNESCO's IdeaLab, Dick Bourgeois-Doyle (2019) summarizes a longer tradition of *two-eyed seeing*—that is, "seeing through the eyes" of one's Indigenous culture and simultaneously "seeing through the eyes" of a majority culture—to bring contemporary AI and Indigenous ways of knowing toward a common framework. These projects provide alternative ethical guidelines that emphasize more traditional Indigenous values of relationship, connectedness, and community.

Ethics of Representation of Data and Results

Ethical challenges can arise in seemingly simple aspects of data science. We will outline two of those aspects here. Please use these concerns *not* as a complete statement-of-concern but rather as a starting point into your own self-inquiry about the ethics of your projects and your protocols. As in our other examples, thinking these problems through will help you to do higher-quality data science work.

Data and results can be represented in many different ways. Some of these may suggest misleading results or obfuscate or detract from important elements of the findings. Different communities may draw different interpretations or implications from the same results. The more complicated or nuanced the results are, the harder it is to represent them appropriately to an audience. Data scientists face a challenge in making judgments about how to represent this information in a way that responsibly reflects the findings of the project.

This leads to storytelling with data, a topic we cover in more depth in chapter 8. The representation of data and results tells a story about the data science problem, the dataset, and the model. Data scientists should try to present this story as transparently and accountably as possible. Data representation is, at its core, a human-centered problem. We should consider the audiences for our work and what they need to understand it. Few students of data science are trained on how to think about the representation of results and data in this way. Instead, much of data science training focuses on the technical development of the model. But a human-centered approach to data science is different. We urge you to be aware of the critical points in the data science cycle where you make choices that shape your data, models, pipelines, results, and representations. Because data science is a cycle and not a linear path, data representation is an iterative process and may lead you to new questions.

Ethics of Training, Validating, and Testing Models

How we construct and test our models is also important for ethical practice and principles.

There are multiple lenses through which we can view the train-validate-test framework that is prevalent in data science courses. Yet the choices that underlie this framework are almost never unpacked. For example:

- Under what kinds of circumstances should one do a 50–50 train-validate split as opposed to a 70–30 train-validate split?

- When is it acceptable to decide that the trained model is validated well on a "held-out" validation dataset but should also be tested on a "never-seen-before" testing dataset?

For instance, the choice of *how* to split training-validation datasets, the choice of *when* to determine that validation is successful, and *how* and *when* the decision is made to test the final model against never-seen-before testing data—all of these choices are crucial. Thinking about the implications of training, validating, and testing is a human-centered data science problem. The best way to address this is to have appropriate background knowledge of the framework and decision points and maintain a freely available written narrative for the decisions that are made surrounding this framework.

Ethics alone is not a solution to all the problems facing people working in data science. Ethics is central to making a good data science project, but it can only provide guidance for your technical decisions; it cannot solve them. For example, ethics alone will not fix a poorly specified model. Ethics may lead you *toward* an interesting and good data science question before you then do the rest of the work, constructing a solid data science method and protocol to address the ethical question that you have chosen to investigate. Ethics is necessary but not sufficient for human-centered data science.

You may face many different ethical dilemmas, and you need to approach each one differently based on the situation. You may need to consult with different parties or organizations, depending on local groups and specific cultural differences. What should be your guiding principles? We suggest that one of your considerations should be the issue of fairness.

Fairness

Is your algorithm fair? This is a difficult question. There are many definitions of fairness, and often people don't agree on the answer (Narayanan 2018). Despite the difficulty of this conversation, we should think it through as much as we can: Are certain groups benefiting more than others? Are certain groups being disadvantaged? Are people with specific demographics more vulnerable to harm by our system? Will our system amplify or perpetuate systemic inequities?

In chapter 7 you will see that there are many systems that attempt to answer complicated societal questions through algorithmic means. For example, many places in the United States use algorithms to decide who goes home on bail and who stays in jail after an arrest. These algorithms were often developed because some people believe that an algorithm will be "fairer" than a human—after all, it is based on statistics and logic and therefore is supposed to be impartial. This is especially important because we know that human decision makers (judges, prosecutors, other people in the position of power) can have difficulty separating their personal feelings from their decision-making and so can (consciously or unconsciously) make decisions that are based on prejudice. People turn to algorithms to help solve these issues, but there are problems with that optimistic view.

First, as we mentioned previously, while many people believe that they know what fairness is, they do not necessarily agree with one another. One researcher listed twenty-one different definitions of fairness (Narayanan 2018). Shira Mitchell and colleagues detailed the derivations of multiple approaches to fairness and documented the tensions among them (Mitchell et al. 2019). Without critically considering these diverse definitions, it is possible to fall into "traps" when creating data science systems (Selbst et al. 2019). In view of these multiple definitions and risks of making mistakes, the problem of formal fairness may be "(im)possible" to achieve (Friedler, Scheidegger, and Venkatasubramanian 2016). These questions are challenging. As one set of researchers observed, "We cannot expect machines to reconcile these differences when society has not" (Rovatsos, Mittelstadt, and Koene 2019, 2). Clearly, there is no "universal" definition of fairness, and therefore it seems premature to trust that an algorithm could compute the "right" type of fairness.

Second, some of these data science systems are proprietary. Their algorithms are not available for inspection (Girasa 2020). Even if we agreed on a "universal" fairness approach, we could not determine whether that approach was implemented in these systems. Because many of the multiple approaches to fairness seem obvious to us, there is reason to worry that the implementers may be writing a version of fairness that "seems fair" or that "makes sense." To whom? On whose authority? And who can evaluate whether their concept of fairness was fairly implemented?

Third, there may be systematic and structural problems with the data in some of these data science systems. A sentencing algorithm that imposes long prison sentences on a group of people will have the effect of removing those people from the free (out of prison) population. The predictive purpose of the sentencing algorithm is to prevent people from reoffending, but we will never know if this group will reoffend because they are in prison and are thereby *removed from the sample* of people whose behavior we can analyze. Our model *prevents them from reoffending* by locking them in prison, and thus makes it impossible to disconfirm the model's prediction. Similarly, a bank loan system is designed to avoid granting loans to people who will probably not repay those loans. However, it systematically denies loans to the people who are predicted to be high-risk borrowers, and therefore there is no way to test the accuracy of the prediction because the high-risk people have been *removed from the sample* of people who received loans. Of course, we cannot know if they would repay a loan that they never received. Thus, these types of systems impose severe limitations on the collection of data that could validate or invalidate their own predictions.

Data scientists need to be accountable for their work and pay attention to the fairness of their algorithms. Accountability means that we cannot wash our hands of the project after we finish the pipeline. We should remain accountable to the people whose data the pipeline is using or whose lives might be changed by its effects. Also, accountability requires that we think carefully and critically about how our pipelines could be used for unexpected purposes. For example, research groups that developed neural-network-based video imitations of famous people did not initially see the potential for harmful misuse of their elegant generative models to produce "deep fakes" (Houde et al. 2020). Yet it is easy to imagine how deep fakes could be used for questionable political purposes or misinformation (Howard 2020). We should be considering what other applications could be found for our work—especially if it can have wide-ranging effects like this case.

Designing Projects for Others to Build On

Reproducibility means designing your projects so that others can test whether they would reach the same results. It is especially a concern for human-centered data science because it takes an approach that considers the people who use or reuse your data/model/pipeline, including possibly yourself at a future time.

- Think about the people who might reuse your data, your model, or your pipeline. Do they have everything they need to reuse them well? What might they need to make your data/model/pipeline better? Who can use your data, and how will they use it? Who can't because they are excluded in one way or another?

Accessibility is another thing to consider. Are people with disabilities unintentionally excluded from building on your work? Does your project work for people who are color-blind or have other vision impairments? We often make visualizations with shades of red and green as the primary colors in diagrams. But those are the hardest colors to distinguish for people with red-green color vision deficiency, which affects about 9 percent of the population.

Openness is another consideration. Can you share your results and how you arrived at them? This is especially important when we consider public-sector projects and working with governments. But this is also important more broadly. If you want your results or pipelines to be useful, people need to be able to access them. For example, for those of us who work within the domain of natural disaster, it is important that people who work in disaster response and relief have access to our results. On the other hand, there are many cases where openness is neither feasible nor desirable—for example, health data or employee data. Thus, there are complex trade-offs between openness and privacy that you need to think through for each project. Consider the benefits when other people can build on the work you have done, and balance that with the potential for harm or loss of privacy.

Readability and legibility mean that it is important to write down your decisions and make them clear, so people don't have to decipher or guess. Logs are a human-centered practice in part because they imagine a future user who will read those logs to understand what the data or the project "is about."

- Have you ever come back to your old code after six months or more? Did you have enough comments to make sense of it? How much time did it take to become "fluent" again in what you did in that code?

Thinking about Your Process and Practice

One key thing that distinguishes human-centered data science is *reflexivity*—that is, actively thinking through your decisions and practices throughout the process. We discussed Donald Schön's work on reflective practice in chapter 3 (Schön 2002) and mentioned in previous chapters that many human decisions go into the data science life cycle. We add here some ways to be thoughtful about those decisions in each step.

We do many things in data science based not on a particular conscious decision but out of habit or convenience. These choices and decisions have an effect on the outcomes of the pipeline. For example, you might not have consciously decided to avoid commenting on your code, but a series of difficult decisions about alternative models may have taken all of your attention. And yet this lack of a conscious decision to prepare documentation will still affect the outcome: other people will have a much harder time using your pipeline and, as a result, may choose an ill-fitting dataset, misuse your models, or misinterpret their results.

Let's start with being reflexive about formulating a question for a data science project. It starts with thinking about how you arrived at a particular question. Did you read what has been done before? Did you consider other studies, possibly from other domains? Have you thought through how answering your question will affect the communities represented in it, or the people whose data you are using? Do you need to work directly with those communities or persons to understand their needs, concerns, and vulnerabilities in relation to your work? Those are all important questions to keep in mind when formulating or reformulating a question for your data science project (Mao et al. 2019). This will help you to make choices based on careful consideration and social responsibility, as opposed to falling into habits or traps of convenience.

As you recall from chapter 2, you need to formulate and document a measurement plan that lays out the steps for how you plan to measure the variables of interest for your question. You should also think through your options for when things go wrong. One likely possibility is that you will not be able to obtain the exact dataset you want to answer your question. If you have to switch to a different dataset, how will you have to change the measurement plan? What kind of bias might this switch introduce into your analysis? Will you have to modify your question, and by how much (and then, is it even the same question)? Thinking through these issues will help you avoid drifting from one question to another to the point of losing track of your original motivation.

- Think of a time when you had to change to a new dataset. How did your way of measuring things change? Did the new dataset map well on your original ideas? If not, how did you need to adjust?

Being reflexive about your data is extremely important but can be difficult, especially if you did not collect the data yourself. You will need to thoroughly consider questions such as: How was the dataset created? Who entered the data? For example, was it the people that the data is about, professionals speaking for individuals, or an automated system? And what kind of organizational power and politics may have been involved? Do you know what the data contains? Do you fully understand what the metadata mean? It is impossible to design a meaningful data science project without a thorough understanding of what each column of data represents, how the variables were measured, and what constraints were placed on the measurement. You might need to contact the people (or organizations) who collected the dataset to get to the needed level of understanding. And then of course: How well does the data relate to your question?

The data practices that shape data help us to think through multiple issues. This brings us to another aspect of thinking through your dataset—namely, whether it was your first choice of data for answering your question. Hopefully, if this was your originally intended dataset, you have concluded that it is the best fit for your question. But if it wasn't your first choice, there are more things to consider: Is this dataset the best *available* fit for your question? Or was it just the most convenient? How are you now planning to answer your question with it?

Being reflexive about your methods is not just the right thing to do, it also leads to better models and better pipelines. Have you thought carefully about what methods you plan to use? What problems might you anticipate with applying those methods? How might they have to be changed or tweaked? For example, you might want to summarize and group social media posts using topic modeling. In data science, the most popular topic modeling technique is called Latent Dirichlet Allocation (LDA), which we first described in chapter 4. But if you think about this method carefully and read about its applications, you will quickly learn that LDA is not very good for working with short documents such as social media posts, and it may not be the right tool for the job. Fortunately, there are other topic modeling methods that work much better for short texts. You might consider all the alternatives and choose the method that is best suited for your question (like the biterm topic model). Are you done? Not quite. The preprocessing you do for the new model might be different, so you'll need to adjust your pipeline for that. The kinds of inputs might be different. And the results produced might deviate from your expectations based on your LDA plan. For example, with the biterm topic model (BTM), you will have to do some extra math to calculate the distribution of topics per document, while LDA

Case Study 5.3
Data Curation at the La Brea Tar Pits: Supporting Data Science by Understanding Data Practices
Andrea Thomer, University of Michigan School for Information

The La Brea Tar Pits are a cluster of incredibly rich fossil deposits located in the heart of Los Angeles—and the home to a unique kind of "big data." An estimated three to four million ice age fossils have been excavated from these deposits since 1901, ranging from microscopic pollen spores to dire wolf skulls to enormous mammoth tusks. These fossils need curation in the traditional, physical sense of "museum curation": that is, each must be cleaned, cataloged, and stored in order to be preserved as part of our cultural and natural heritage and to be used in future scientific investigations. Furthermore, the *data* associated with each specimen needs to be curated as well. These data range from field notes documenting each fossil's original location in the deposit, to databases indexing the site's massive collections, to detailed anatomical drawings and computed tomography (CT) scans used in evolutionary biology studies, to protein sequences and radioisotope measurements derived from the fossils. These digital objects need as much care as the physical fossils to be used in data-intensive science and to ensure that they are accessible by future generations.

Data curation (also called *digital curation*) is the work of making data usable, sharable, accessible, and preservation-ready over its lifecycle of interest to science and scholarship. This work—and the people who do it—are foundational to data science, but often overlooked. Data curation includes cleaning, reformatting, annotating, and standardizing data for analysis; describing data for later retrieval or reproducibility; and taking steps to ensure that data is stored in a stable format and trustworthy repository. Data curators go by many aliases: data librarians, research data managers, data stewards, data janitors, data engineers, and more. They also hold diverse roles in research teams: working directly with scientists as part of a lab, working in data repositories, or consulting from posts in an academic library, or anywhere in between. Regardless of their name or position, though, they share a common cause of making data *fit for use* and accessible to broad audiences now and in the future.

The work of data curation is rarely one-size-fits-all. Curation must suit the intended use of a dataset and the organizational context, and it must take into account existing *data practices*—the workflows, cultures, and moral economies (i.e., the ethical values governing the data "marketplace") of data use in a community (Strasser 2006). At La Brea, this entails supporting both very old and very new ways of collecting data. Scientists at La Brea have collected specimen data in the same way since 1969, and the data collected via this legacy workflow needs to be supported going forward. However, they must also manage data collected with novel methods, such as those from recent studies of "food webs" (the networks of predator-prey relationships in an ecosystem) at the site. While the curators at La Brea have recently invested in a new collections management system, there are limits to what kind of data it can store, and they have had to augment their database through *ad hoc* catalogs stored in spreadsheets. There will likely never be one database that will magically "solve" their data curation needs, but rather, a rich sociotechnical ecosystem of curatorial systems and workflows.

Studying and understanding data practices isn't just important for the development of effective data curation protocols. Understanding data practices is also critical in surfacing the cultural norms, assumptions, biases, accidents, and perspectives that go into a dataset's creation. Because of the unique data collection method used by La Brea researchers, a dataset collected at La Brea includes variables and details that likely wouldn't be collected at another paleontological site. This doesn't mean one dataset is better or worse than the other—but rather, that the *people* who collected each dataset chose to record different data points and therefore emphasize (or obscure) different aspects of their study sites. Thus, studying data practices can help shed light on the ways that human choices shape data that may otherwise seem "objective" or "value neutral."

automatically does that (however, bypassing this step is exactly what makes BTM more suitable for short text).

- Think of a time when you had to pivot from one method/model to another. How much work was entailed in preprocessing? What about lining up the model inputs with your pipeline? How different was the format of the results?

Our goal here is to make you aware that thinking critically about your process and how it affects your results, as well as the people whose behavior is represented in your dataset, is needed every step of the way. Thinking through your process may be even harder when some parts of your system are automated. In this case, you have less control over that part of your pipeline, as it may feel like an opaque box. We still encourage you to be as reflexive as possible about these parts of your pipeline by getting enough information about the automation to understand exactly what it does and how. Otherwise, you will be importing decisions wholesale, and you might be surprised by their outcomes in terms of whom they affect and how.

Platform Affordances and Data Schema

In thinking through your process, there is another source of potential trouble. Because so much of the data used in data science projects comes from various online platforms, we need to highlight the platform affordances and how they affect data schema.

Affordance is what the environment offers the individual. The term originally comes from psychology, but it was appropriated by people working in human-machine interaction. In that new context, affordances are the possible actions that technology makes available to someone based on how they perceive them in their environment (Nagy and Neff 2015). Here is an example from the realm of online platforms. Until a few years ago, the only way you could react to a friend's post on Facebook, outside of writing a comment, was to click the Like button. Clicking on the button was easy but not always very meaningful. If your friend shared some sad news and you wanted to show that you were paying attention and being sympathetic, a *Like* probably did not literally mean that you liked the bad news. But the affordances of the platform put you in a difficult position. How would your friend interpret your use of the Like button? Here the Like button is an affordance: it enables certain visible actions and constrains our space of possibilities for others (such as expressing more nuanced emotions). Understanding this, in 2016 Facebook introduced its Reactions feature, broadening the type of emotional reactions we could signify with a button click—and so broadening the affordances of its interface.

As you might imagine, the affordances of an interface—what you can easily do within online platforms—translate into the data we can collect from these platforms. Now, instead of one column documenting how many *Likes* each Facebook post got, we would have six different columns showing how many reactions of each type it received. Hence, the affordances of the platforms can be traced directly to the data schema of the data that we collect from them: how the data is organized, what columns are present, and what they measure. And, of course, this determines the kinds of questions you can answer and the kinds of analyses you can run with this data. It is important to think carefully about what the data schema from the platforms affords you—that is, what it allows you to do or restricts you from doing—as a data scientist.

How data is packaged, stored, and distributed makes certain analyses easier than others. For example, data you can get from various Twitter APIs is organized around individual tweets. At first glance, this is a reasonable organization. But as we will describe in more detail in chapter 6, people use the system conversationally. They refer to their own previous tweets (as you would in a conversation, expecting that others remember your previous statements) and to tweets by others (as you would build on what the other person said in a dialogue). From this perspective, isolated tweets are not very useful, and we need whole Twitter conversations to understand and model what people are saying. And yet very few data science projects are doing that, because reconstructing those conversations from isolated tweets is a lot of work. It's much easier to rely on individual tweets, which are directly available from the Twitter APIs.

What is directly available in the data schema greatly affects the types of analysis that are common. We all know how easy it is to pick the path of least resistance, to look under the streetlight. This means that we often choose the same easy proxies for the things we want to measure for our questions. And of course, these are not always the best or the right ways of measuring those concepts and answering our questions. So just like with other parts of the data science pipeline, we have to be very thoughtful about how the data schema affects our analyses.

- Think about a time you chose a concept to analyze because there was a column for it in the dataset. Were there other, better ways to answer your question? Were they more labor-intensive? How were your results affected by this choice?

Many people use social media data in data science projects. There are so many sources of information and so many loud voices there. Which ones do we attend to? Attention on social media is easiest to measure through the traces of what people do with the content: when they click to retweet, favorite, reply, or mention. Those metrics are directly available in the data schema, and they are easy to count. But they are also completely dependent on social media users taking action—pressing a button in the interface. What about measuring the attention that we pay to posts we just read, without a mouse click? If no action was taken, there is no digital trace, so this information is not recorded in the social media data. To measure that kind of attention, we would have to go beyond the easily accessible data from platform APIs to capture what people are doing through other means: eye-tracking studies, direct observation, or interviews. Of course, these other methods are more labor-intensive and time-consuming than working with existing datasets.

Carefully thinking through the constraints of the data schema, and how they affect your analysis choices, is an important part of being reflexive about your process. People in data science often use the following techniques for reflecting on their practice:

- *Ask others* in data science how they do their work and how they think about their work.
- *Observe others* as they do data science work and as they work with demos.
- *Ask* data science workers to lay out their work practices in detail, and to explain those work practices in detail as they lay them out.
- *Analyze code and documentation* that others write—or do not write—as they do their work.
- *Read* the online descriptions and instructions at popular sites that collect user-provided data.

Conclusion

In this chapter we introduced you to many things you should consider when designing and implementing your data science project. These things are what distinguish human-centered data science: formulating a meaningful question, considering issues of ethics and fairness, designing projects that others can easily build on, and incorporating reflexivity into your entire process. Many of these suggestions are concerned with people: the people represented in your dataset, the people whose lives may be affected by the results of your analysis, the people who might want to reuse your pipeline or just your results (such as classification labels), and even yourself at a later point in time. These strategies make your data science project more human-centered and sensitive to the people who are a part of practically every step of the cycle.

These strategies also make for a better data science project. Formulating an interesting question that matters to others and can be tested with well-suited data will make for a powerful finding. Carefully considering the constraints of the data schema lets you refine and answer more nuanced, more powerful questions. You will get stronger results because you are deliberate in what you are trying to measure when you have carefully planned your measurements. Thinking through your model, including considerations of fairness, will produce more precise and generalizable results and pipelines.

Ethics, like reflexivity, is not a one-time consideration. That is why recipes and checklists don't work in this domain: we never know when or what parts of our work will require careful and critical consideration. We emphasize reflexivity as a practice—something you do (and continue training yourself to do) at every step of the process. It is a skill that gets easier with time, as long as you are open to asking yourself difficult and critical questions.

Recommended Reading

Bourgeois-Doyle, Dick. 2019. "Two-Eyed AI: A Reflection on Artificial Intelligence." Canadian Commission for UNESCO's IdeaLab. https://en.ccunesco.ca/-/media/Files/Unesco/Resources/2019/03/TwoEyedArtificialIntelligence .pdf. This paper describes how members of an oppressed culture or group may need to maintain two ways of perceiving social realities: one in terms of the truths of their own culture, and the other in terms of the normative beliefs of the oppressing culture.

franzke, aline shakti, Anja Bechmann, Michael Zimmer, Charles Ess, and the Association of Internet Researchers. 2020. "Internet Research: Ethical Guidelines 3.0." https://aoir.org/reports/ethics3.pdf. This extensive ethics guide for responsibly using online and social media information was developed by an international organization of academic researchers.

Quinton, Sarah, and Nina Reynolds. 2018. *Understanding Research in the Digital Age*. London: SAGE Publications. This general guide to digital research is organized around the themes of ethics, expectations, and expertise. The book provides a way to think through a research project that focuses on digital data.

Smith, Linda Tuhiwai. 2013. *Decolonizing Methodologies: Research and Indigenous Peoples*. London: Zed Books. This work helps you to reconsider your own cultural frame as it influences your data science method. While not about data science as such, this book is very much about how we see other people—how we categorize them and how we make them visible or invisible in our analyses and our reports. Although it focuses on colonized peoples, its lessons apply to any situation that involves two or more groups with different degrees of social power.

Tufte, Edward R. 1983. *The Visual Display of Quantitative Information*. Cheshire, CT: Graphics Press.

Tufte, Edward R. 1990. *Envisioning Information*. Cheshire, CT: Graphics Press.

UK Government Digital Service. 2020. "Data Ethics Framework." https://assets.publishing.service.gov.uk /government/uploads/system/uploads/attachment_data/file/923108/Data_Ethics_Framework_2020.pdf. This is an example of how a national-level strategy for the ethical use of data might be implemented. This framework document is also useful for its clarity and ability to explain data science ethics questions in practical and applied terms to those outside the data science profession.

6

Human-Centered Data Science Methods

This chapter outlines key traditions from social science, design studies, and critical theory to show other ways to study human behavior and how they could be incorporated into or combined with data science. The results of these combinations and experiments at the intersections of data science are quite exciting as they merge computational capacity with our abilities to interpret, build, and transform the social world we live in.

Social Science Methods for Rethinking Data Science

Emile Durkheim, a French philosopher, is credited with founding the field of sociology at the end of the nineteenth century. He worked at a time when society was undergoing enormous changes. In Europe, the rise of manufacturing and the use of more machines on farms meant many people were forced to move from close-knit villages to relatively more anonymous cities. Traditional village values were giving way to more connected, cosmopolitan, and global world views. Durkheim suggested that there were what he called "social facts," ways of acting, thinking, and feeling that are influenced by the communities we live in. These social facts, Durkheim hypothesized, have some influence over us (Durkheim [1895] 2014).

In 1897 Durkheim published *Suicide*, one of the first books that used data to explain the behavior of people in society. Durkheim wanted to test theories about social cohesion as a social fact. How much social cohesion was needed to hold society together? How much was too much? The problem was that social cohesion and social integration were difficult to measure. Durkheim's idea—that cohesion was an attribute of groups instead of being an attribute of individuals—was interesting. But he couldn't *see* social integration, so how could he measure it?

He did it by explaining patterns in variation, and building those patterns into larger explanations, not unlike the process of modern data science. However, Durkheim painstakingly hand-calculated death rates using death certificates from different cities and countries. In this data, Durkheim found sets of patterns: substantial differences between men and women, between Catholic and Protestant communities, between city and countryside. People who underwent major life changes and those who lived in rapidly changing places were more at risk for suicide. Men whose wives had recently died were more likely to commit suicide than women whose husbands had died, and they were more likely to commit suicide than men who had never married. Older people were more at risk for suicide than younger people. These patterns all suggested that a third variable—changes in life circumstances—had some significant influence on a person's tragic, seemingly deeply personal choice. If suicide were simply a function of individual psychology, these patterns would be difficult to explain. Durkheim used these differences to develop a hypothesis about how people are influenced by the rules of everyday life. Too many or too few rules, or too much or too little integration with others, put people at a greater risk for suicide. Durkheim used simple tables and charts to map these differences in variation and to create a theory about the role of social rules on people's lives (Durkheim [1897] 1979). His data science tools were rudimentary, but the spirit of his approach to demographic data is very much in keeping with the spirit that motivates data science today.

Social science has over a hundred years of experience with studies to explain human behavior. Our goal in this chapter is not to convert readers into social scientists. Rather, we want to show how social science and other methods can bring a human focus to data science, while pointing interested readers to the resources they need to dive deeper if they wish. We introduce a range of methods so that these approaches can be considered in designing, evaluating, and feeding back into the data science pipeline. They can be thought of as ways to inform projects with the contextual information that is often missing from large-scale data; additionally, the information may help you to find and understand social science colleagues if and when you need them in your work.

Data Science and Context: Why This Matters

Human behavior depends on *context*, by which we mean information about users, groups, situations, and environments that are outside the data but associated with the data or its generation in some way. The association may be historical (e.g., you need to know what happened before the data was collected), cultural (e.g., you need to know what this word means for this population or profession), or associational (e.g., you need to know the social connections of this person).

Context also covers how people behave differently in different situations, such as the difference between weekdays and weekends, between home and work, and so on, depending on the situation. Context matters for people's daily routines. How people shop for groceries may be dramatically different before and after a major storm, pandemic, or other large-scale societal event; it is influenced by, for example, the times when people believe that grocery stores will be open and stocked and a shared awareness of which items are in high demand. The COVID-19 crisis showed these kinds of phenomena very clearly in bulk-shopping for toilet paper and hand sanitizer. To get data science right, we need to understand the specific context of what people were doing, what they were thinking, and the factors that may have influenced their behavior. Knowing when the data originates and contextual factors that may have been at work are crucial to running appropriate analyses, utilizing the appropriate models, and gaining the most insight from the data. As mentioned in chapter 3, an influential approach called Value Sensitive Design (VSD) analyzes context in terms of conceptual, empirical, and technical investigations (Friedman, Kahn, and Borning 2008); we also address this in chapter 7.

Context-aware computing and context-sensitive recommendations are increasingly important, and often what people mean when they say context is relative. The problem with context, anthropologist Nick Seaver (2015) has argued, is that "everybody has it"—context is recursively defined as information outside measurement. Seaver studied music recommendation algorithms that rely on getting users' "context" right: users may want very different music when they are listening alone at home versus working out at a crowded gym. Context is "simultaneously missing from data science and central to it" (Seaver 2017, 1103). For data science, one problem is representing context in the data, which we'll discuss below. One criticism of data science, and quantitative studies in general, is that these methods can miss the context that makes results interpretable, understandable, and important. Context can be centrally important when working with communities, as one research team noted: "When designing technologies for social justice, we must ensure we adequately understand the contexts in which we design, including but not limited to the social, historical, political, and legal circumstances" (Strohmayer, Clamen, and Laing 2019).

For example, people can recognize something as an instance of irony or sarcasm depending on the situation and on their sense of humor. People have different backgrounds and experiences that affect how they act and what they perceive about others. Someone who grew up, for example, in the former Soviet Union will have a drastically different history with the concept of socialism than someone who grew up in the United States. Concepts of appropriate relationships to land are very different between European-descended Americans ("A place I bought today") versus Indigenous Americans ("Our only place in the universe"). The same is true for concepts about data, which may be quite different between European-descended perspectives and Indigenous perspectives (Duarte et al. 2020; see also Maitra 2020; Nunavut Department of Education 2007). When examining a less powerful group encountering a more powerful group, a form of "two-eyed seeing" (as described in chapter 5) may be necessary to maintain a simultaneous awareness of both cultural contexts (Bourgeois-Doyle 2019)—that is, seeing honestly from the minority culture's perspective ("What is our reality?") and seeing defensively from the majority culture's perspective ("What do they think of us?").

Our backgrounds, the experiences we have throughout our lives, the people we know—all inform how we view the world and shape how we behave in it. Linda Tuhiwai Smith (2013) analyzes "how we view the world" in formal terms as the *frame* that we put around some situations or people. Those differences are likely to influence the data in profound ways. And yet such differences are probably not visible in the data. As Marisa Elena Duarte and colleagues wrote, "Of course, data can never fully represent reality" (Duarte et al. 2019, 163). These differences may be especially influential in datasets concerning social interaction, where the contextual factors for multiple people intersect. That is why it is important for data scientists wanting to ensure human-centeredness in their projects to account for the social context of their data.

Knowing the context is important for understanding the background of the data and interpreting data. But context differs by how and what was measured in the data. People doing data science can make choices that offer context as additional structures to be utilized by models and algorithms. For example, if we know that there should be some temporal shifts and changes in the dataset, we can use that knowledge to add one more constraint to the model or provide it with more structure and information. Katharina

Reinecke collected a very large dataset of scheduling activities in Doodle, a popular online scheduling tool, and found that national culture can be an important contextual variable in studies of both time perception and time-oriented behaviors (Reinecke et al. 2013). In another example, we may need to know whether each person in our data sample lives in an urban, suburban, exurban, or rural culture.

Thick Data and Its Importance for Data Science

As a consultant to Nokia, Tricia Wang noted that her in-depth studies of mobile phone users in China were having a hard time getting heard by decision makers at the company's Finnish headquarters, who were proud of their reliance on data science. Wang saw from her research that lower-income users in China were ready to pay more for smartphones, but Nokia's business model was built on providing them primarily for high-end users. Her findings came not from the majority view of large studies of users but from the outliers and edge cases. She demonstrated this with an in-depth look at 100 users in China. She argued, "What is measurable isn't the same as what is valuable" (Wang 2016). Nokia missed what its competitors saw: that more people would use the features of phones as computing power increased. The experience led Wang to start her own consultancy, give a TED Talk, and increasingly advise companies on why "big data" needs "thick data," drawing on the concept of "thick description" in anthropology (e.g., Geertz 1973). Thick data, like thick or "rich" descriptions, draw on deep insights from a few people over "thin" insights from large-scale quantifiable data. In her words:

Thick Data is data brought to light using qualitative, ethnographic research methods that uncover people's emotions, stories, and models of their world. It's the sticky stuff that's difficult to quantify. It comes to us in the form of a small sample size and in return we get an incredible depth of meanings and stories. Thick Data is the opposite of Big Data, which is quantitative data at a large scale that involves new technologies around capturing, storing, and analyzing. (Wang 2016)

Wang's approach has had a significant impact in industry and on the field as a whole. We suggest that a human-centered approach to data science does not see opposition in "thick" and "big" data. Others agree. The UK government's Policy Lab has articulated how thick data and big data might work in government data science projects for the design of better government services (figure 6.1).

How can we combine big data and thick data? What types of analyses and processes lead data science to the "meso" or in-between focus between people and their societies, as shown in figure 6.1? Researchers are still working on how "big" and "thick" analyses may be constructively used together. Will topic modeling and qualitative research eventually converge toward similar or equivalent conclusions (Baumer et al. 2017)? Or can we imagine how cycles alternating between big and then thick data could inform and shape the next analysis in the cycle (Muller, Guhaet al. 2016)? Regardless of which path informs practice, data science may benefit from combinations and experimentations with such methods.

Below, we first look to fields such as social science and design to think about how to bring human-centered approaches into data science, describing quantitative and qualitative social

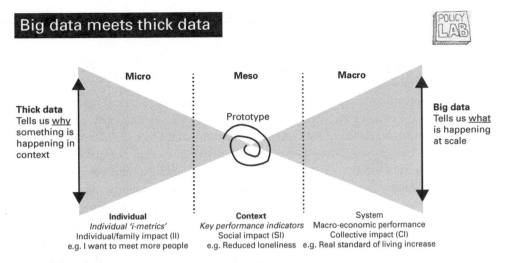

Figure 6.1
UK Policy Lab's model for combining big data and thick data (Siodmok 2020).

science methods, before turning to what are known as transforming, critical, or interventional methods from both design and social science. We then look at how human-centered data science methods can expand existing data science techniques by taking advantage of the rich context present in the data.

Quantitative Social Science Methods

Quantitative social science methods are anchored in a theory of knowledge that sees the world, like Durkheim did, as consisting of social facts that can be measured or known through careful, testable ways of observing phenomena.

Quantitative methods analyze numerical data using mathematical or statistical approaches. In the social sciences, the term *quantitative methods* groups together many different types of data sources (surveys, documents) to focus on the data type (numerical) and the techniques of analysis (primarily statistical). Types of quantitative methods include the analysis of survey data, the analysis of documents through quantitative content analysis, and the analysis of data from experiments. Quantitative data can also be obtained from collections of documents, such as extracting quantitative medical test outcomes from electronic health records or house sizes, costs, and attributes from real estate listings.

Quantitative social science research shares some common traits. The research is often designed with a clearly defined research question at the beginning of the project. Data takes the form of numbers and statistics and is presented in nontextual form. The goals are to *generalize* from the study—that is to say, to use the results to say something about the concepts on a wider scale, predict future outcomes, or show how two or more concepts relate to one another. The results are based on samples that represent a larger population, such as 1,000 likely voters in the next election surveyed to reflect the opinions of people who are likely to vote.

See the recommendations at the end of this chapter for more in-depth coverage of these methods.

Computational Methods in the Social Sciences

Although quantitative social science has a long history, there are also newly developed (and developing) social science methods that take advantage of new computational techniques and are relevant to human-centered data science. *Computational social science* leverages the relatively new capacity to collect and analyze data with unprecedented breadth, depth, and scale and draws on existing social science strengths in explaining human behavior, moving toward more scalable methods to analyze complex social phenomena (Lazer et al. 2009). *Social data science* refers to using theories, concepts, and methods from both social and computational sciences "to analyse unstructured heterogeneous data about human behaviour, thereby informing our understanding of the human world" (Oxford Internet Institute 2020). The social sciences have been better equipped for understanding and accounting for context, especially in the case of qualitative methods (described in more detail below), where researchers attempted to deeply understand people's perspectives by either observing them for a long time or conducting in-depth interviews with them. Qualitative methods are powerful for understanding people's perspectives, but also highly labor-intensive and hard to scale to large groups of people. There are many excellent examples of computational social science work. Here are a few that combine and take advantage of multiple methods from academic, government, and industry researchers:

- Researchers used location data from Wi-Fi, Bluetooth, and GPS on 100 mobile phones to infer who was friends with whom and to what degree. They compared this data to the answers that people gave in surveys to test whether cell phone data could accurately predict friendship status. This was one of the first studies that advocated for computational social science as a methodological approach to answer complex social science questions (Eagle, Pentland, and Lazer 2009).

- Researchers analyzed hundreds of millions of Chinese social media posts and found that the Chinese government hires about two million people to post about 448 million comments a year to shift the conversation on social media toward positive opinions of the Chinese government (King, Pan, and Roberts 2017).

- Researchers used a validated questionnaire for measuring depression to test if people's public tweets could be used to find words that might be indicators of depression as an early-warning system for potential clinical intervention and support (DeChoudhury et al. 2013).

- Two of the authors of this book (Shion and Michael) used a combination of company-wide surveys and social media to understand patterns of emotional "contagion" of IBM employee attitudes (Guha et al. 2016; Muller, Shami et al. 2016). They found that employees were most influenced by people in the same group, followed by their friends in the company social network, followed by managers. One year later, a more statistically complex study by Tanushree Mitra and colleagues found that a positive emotion (employee engagement) came largely from employees' peers, whereas a negative emotion (disengagement) came primarily from managers (Mitra et al. 2017).

Qualitative Social Science Methods

Qualitative social science methods focus on *qualities*: meanings, values and processes, rather than easily measured amounts. Qualitative research aims to capture how people make sense of their own situation and to reveal the context where they do so. Qualitative analysis can refer more broadly to the analysis of nonnumerical data. These methods focus on how people experience and give meaning to the things that happen to them. Qualitative social research can rely on researchers' disciplined, rigorous interpretations (Charmaz 2006; Clarke, Braun, and Hayfield 2015; Corbin and Strauss 2014; Muller 2014). Qualitative research reflects people's multiple subjective experiences and standpoints, rather than a single objective or generalizable depiction of social facts. This means that qualitative social science methods are good at capturing insight and information that adds context to large-scale data and that can help human-centered data scientists increase the explanatory power and contextual richness of their investigations.

Because qualitative research relies on interpreting people's actions and behaviors, it is not necessarily easy to do. Imagine that you are a data scientist with a mobile app company. Your company has near-perfect data on the actions that users take within the app. But without understanding the context of those choices or what those choices mean to users, you lack insight on why they make those particular choices and what that might mean for the company's strategy.

One method that may help is *observation*, watching and noting how people do particular things, while producing careful and systematic notes, known as field notes. Another method is *interviews*, asking directly, formally or informally, using either structured or open-ended questions. *Focus groups* are a type of interview with several people at once to get their reactions and responses and to use how they respond to one another and think together to shape the results. *Participatory analysis* invites people to construct a description of their experience (e.g., through text or diagram) and then to interpret their own description; both description and interpretation become data for analysis (figure 6.2). The data produced from qualitative research could be reported as stories or *vignettes*, such as in-depth user profiles, interpreted holistically or thematically, or analyzed quantitatively.

Ethnography is a technique rooted in the fields of anthropology and sociology that involves the in-depth exploration of participants' experiences in a particular culture or social situation. The technique is used to understand specifics of people's behavior and explore the underlying, perhaps unspoken, reasons for their behavior as individuals and groups. In the early history of ethnography, Western anthropologists typically traveled to and lived among different cultures and communities to report on their social systems (with obvious issues of bias and power imbalance). More recently, social scientists, researchers, and data scientists have been using ethnographic methods to understand more deeply their own society, including how people interact with technology and data. Several of the authors of this book (Cecilia, Gina, and Michael) have used *digital ethnography* and participated in online communities to observe behavior that occurs in virtual spaces (Aragon and Davis 2019; Muller and Chua 2012; Schwartz and Neff 2019).

Qualitative methods take time and are difficult to scale. The payoff, though, is that qualitative research adds insight about a particular experience, emotion, thought, or cultural background that may be challenging to capture otherwise. For an example, let's go back to

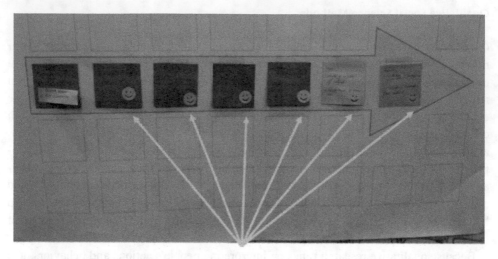

Figure 6.2
Participatory analysis of one person's experience with the data science cycle. The faces are decorative stickers that indicate steps in the cycle where collaboration took place. In addition to the stickers, the informant provides rich contextual information about why the project was done, who was on the data science team, and how their work practices contributed to the outcomes of their project.

thinking about data science for a mobile app company. We could ask users questions about the app to help give context and insight to a finding from the user trace data. We could host a focus group to see how a group of users respond to a change that was suggested from the large-scale data. We could watch as people navigate the site to see why certain parts of the app were less accessed than others. Usability researchers have developed principled ways to do this, such as *heuristic evaluation* (Nielsen 1992) and *cognitive walkthrough* (Lewis et al. 1990). This method has been applied and extended to study how mobile apps frame users' self-expression, relationships, and interactions by looking at the vision, operating model, and modes of governance for an app and tracing systematically and forensically various stages of registration and entry, everyday use, and discontinuation of use (Light, Burgess, and Duguay 2018). These are examples of *elicitation* techniques that use a picture, a visualization, or a device to elicit information from someone; researchers have also used results of data science and visualizations to get feedback from people.

So far, we have talked about qualitative social science research as an approach to collecting data. Qualitative research also entails ways to analyze nonnumerical data. *Grounded theory* (e.g., Charmaz 2006; Corbin and Strauss 2014; Muller 2014) and *thematic analysis* (Clarke, Braun, and Hayfield 2015) are two methods used for interpreting themes in qualitative data, with a more than fifty-year history since the originating work by Glaser and Strauss (1967). While they have some key differences, both grounded theory and thematic analysis provide ways to help the researcher discover and interpret themes in qualitative data and to *build theory* from the data in a rigorous and (qualitative) data-driven way.

Transforming Methods: Design and Critical Approaches to Data Science

Design and critical methods come from a *pragmatic* tradition that sees intervention in the world as the goal. These methods have relevance to data science in part because design has a close connection to computer science, for example, and human-centered design is a long tradition in technology design. Critical approaches aim to intervene in the world, starting from a values-driven (Friedman, Kahn, and Borning 2008) or social or political position (Asad 2019) of wanting to do what is right, just, or good. These approaches acknowledge that these values are not universal or objective but very much depend on the person (Alsheikh, Rode, and Lindley 2011; Borning and Muller 2012; Le Dantec, Poole, and Wyche 2009). These *transforming* methods seek to offer intervention in the world to effect change. Transforming methods see knowledge creation as part of recreating the world, not just objectively describing it (D'Ignazio and Klein 2020; Haraway 1988; Harding 2004; Iliadis and Russo 2016). In design and critical methods, people's individual experiences are seen as legitimate sources of knowledge about the world, and the commitment to making change in the world is as important as seeking truth or doing science.

In many of these transforming methods, critique is a technique for changing and improving the thing we are studying (Iliadis and Russo 2016; Kitchin and Lauriault 2018; Neff et al. 2017). Critique is a way to identify the pain points, challenges, and problems to help understand what to improve. For example, as part of the work that one of the authors (Gina) did on data science for workplaces, she and her collaborators proposed a way to understand the data science pipeline on work by asking what and whose goals are being achieved or promised through what work, done by whom, under whose control, and at whose expense.

Action research works with communities with the goal of having impact on the world according to a plan that is agreed on among the community and the researchers. Action research recognizes the enormous power and responsibility that data science has in remaking and reshaping how we know the world, as well as its potential in how communities and people can be empowered to ask questions.

If the goal is to change the world, then we may need to ask a series of "who" questions. A feminist analysis might ask, "Who gets to change the world?" An activist might ask, "Based on whose authority do *you* want to change *our* world?" In critical approaches, *someone* is doing the critique—alone or with partners or allies (Feinberg 2007).

Anthropologists distinguish between two types of descriptions of a situation: *etic* descriptions are written from an external observer's perspective, while *emic* descriptions are written from the perspective of a person who is in that situation. There is value in outside, or etic, critique as applied to data science. Sometimes an outside observer can ask questions that would not occur to someone who is deeply involved inside a workplace or a life process from the emic point of view. Sometimes people inside the situation can offer their own emic critique. These two sources of critique can become more powerful when they join forces.

However, outsider or etic accounts have been criticized for their imbalances of social power. There is a problem with speaking "for" others when an observer or researcher in a powerful position becomes the reporter about the work or lives of less powerful people (Alcoff 1991). Finding partners and working with communities to create a joint critique can be more powerful and insightful because it combines both insider and outsider knowledges and perspectives (Muller 1997; Muller and Druin 2012). Participatory design emphasizes "joining forces in design" (Kyng and Mathiassen 1995). Action research provides a different set of practices through which a community and external researchers can work together to develop knowledge and agendas for change that are based on the community's needs (Hayes 2011).

Human-centered design means designing something—whether it is a system, a web-site, or a social policy—with the people who will use it in mind and with sensitivity to the communities of people who are affected by the process. One of the authors of this book (Cecilia) used human-centered design to work closely with astrophysicists to develop software that would allow them to do their work of studying supernovas more effectively (Aragon, Poon et al. 2008). A hospital designing a new children's ward might want to consult children on what is important to them, as well as parents and the people who will work in the new space. To get at these different views, human-centered designers will often use the tools of social science.

Participatory design goes beyond consulting to involve people in the choices and designs of the artifacts that they will use (or that will be used on them). Michael used participatory design at a software company to engage directly with users in the design of new technolo-gies to support their work (Muller and Carey 2002). Over the years, researchers and practi-tioners of participatory design have developed diverse theories and methods that range from serious to playful, from prototyping with words to prototyping with low-tech materials to prototyping with software and hardware (Muller and Druin 2012). The most important aspect is that the people who will use a system or service become co-designers of that sys-tem or service, combining their work knowledge or life knowledge with other knowledges of technical and design professionals. These approaches enrich the design teams' knowledge and simultaneously enrich the critique through combinations of diverse stakeholder perspec-tives (Bødker et al. 1988).

Design fictions are used to create narratives to think through possibilities. A design fic-tion is a piece of speculative storytelling that uses fiction to enable deeper exploration of the potential implications of design elements. For example, one of us (Michael) worked with colleagues to understand the potential for both substantial benefit and harm from new generative algorithms in data science. The team became worried about potential mis-uses of these new technologies to create false "evidence" in law or insurance cases or to reuse a writer's style algorithmically in works that are not attributed to the writer. Using strategically incomplete stories, they asked their informants to finish writing the stories to consider (a) whether these were actual misuses and (b) how individuals or societies might respond to these new challenges (Houde et al. 2020). Design fictions can be used in sev-eral ways. They can be presented as complete narratives with interpretive commentary to warn of possible futures (Fiesler 2019), or they can be used to elicit commentary and cri-tique (Sorell and Draper 2014). Design fictions can also be used as participatory openings to collect and analyze narratives from stakeholders in future technological designs (Cheon and Su 2018; Muller and Liao 2017).

While some design fictions may seem straight out of the British dystopian sci-fi show *Black Mirror* (a great example of design fiction), they highlight how important it is to think through the implications of data-driven projects. The Ethical OS toolkit (https://ethicalos.org/) uses design fictions to create scenarios to help people think through the ethical implications of the technologies they are building. As an example, the Ethical OS project asks people building new technologies what the implication for their product or service would be under several scenarios, such as bank data combined with social media data to determine credit scores or blockchain tools used to prove sexual consent. To explain the reasons for this work, the project's authors write that "if the technology you're building right now will someday be used in unexpected ways, how can you hope to be prepared? [Which] choices can actively safeguard users, communities, society, and your company from future risk?" (Institute for the Future and Omidyar Network Tech and Society Solutions Lab 2018). We suggest people in data science ask themselves these questions, and we have described a range of methods that you can use—by yourself, with colleagues, and/or with current or future users.

Design justice seeks to use design to rectify injustices or inequalities. For example, following the principles of Inuit Qaujimajatuqangit, Inuit ways of knowing involve observational learning and respect for the collective knowledge of elders (Nunavut Department of Education 2007). This view of cultural knowledge contrasts with the individualistic achievement promoted by European-descended educational systems. Sasha Costanza-Chock (2018), borrowing a phrase from the Zapatistas, calls for "a world where many worlds fit." Mariam Asad (2019) and Angelika Strohmayer and her coauthors (2019) take related positions when they propose that design be done so as to enact the future that we want to live in. The challenge for data science is to analyze our data so that people from multiple perspectives or standpoints can see their realities represented in our analyses.

Critical race approaches look at how data might intersect with racial inequalities in society. When her then-young daughter searched online for "Black girls," Safiya Noble found that she was exposed to highly sexualized content. But these searches returned very different results for "American girls" or "white girls." Seemingly neutral search engine results may reflect deep social biases (Noble 2018). One of the authors (Cecilia) had the same experience when she searched social media for the hashtags #Filipinas or #Latinas. As an MIT graduate student, Joy Buolamwini found that commercial facial recognition software was significantly more accurate for classifying images of people with light skin compared to images of people with dark skin (Buolamwini 2016; Buolamwini and Gebru 2018). Ruha Benjamin calls this the "New Jim Code," the use of "new technologies that reflect and reproduce existing inequalities but that are promoted and perceived as more objective or progressive than the discriminatory system of a previous era" (Benjamin 2019, 1).

Critical data studies looks specifically at data's role in supporting power within society: how data is generated and curated and how it exerts power (Iliadis and Russo 2016). Research in critical data studies asks big questions about the roles of politics and economics in data and frames the questions in terms of critique: "Should we do this analysis?" "What will be done with these results?" "Who does this project benefit?" These broader questions can never be directly answered by the data, but critical data studies can offer much to data science by providing a way for these kinds of reflections to happen in data science teams (Neff et al. 2017). Such questions teach us to think through where data comes from, the power imbalances represented in the data, how data is made, and who the data benefits.

The *data justice* movement looks at questions of "fairness in the way people are made visible, represented and treated as a result of their production of digital data" (Taylor 2017). These approaches are needed, as one group proposes, because data has the power to shape what kinds of information are valued, what is knowable, and what is acted on (Dencik et al. 2019).

To see how data justice might influence a project, consider the metadata of mobile phone records. One team whose work was published in *Science* showed how the data could estimate income levels in Rwanda, a country with very little good economic data (Blumenstock, Cadamuro, and On 2015). Such analysis might help those who work in economic development. However, imagine government efforts to identify the characteristics of particular people—say, refugees or people illegally crossing national borders. Values about border security or humanitarian aid will shape the data scientists' view about whether this is a "good" project to work on. Data, Linnet Taylor writes, has "the power . . . to sort, [categorize] and intervene" (Taylor 2017; see also Bowker and Star 1998). Data justice approaches teach us to ask who is wielding this power and to what ends. During the COVID-19 pandemic, many governments explored policies to put social and location-monitoring software into people's mobile phones to help track the spread of the disease. The initial purpose for early detection of possible contagion of the novel coronavirus was benign and even altruistic. However, data justice approaches ask if there could be less benign uses for this very intimate data.

Like data justice and critical data approaches, feminist approaches reflect a view that one's position or standpoint in society shapes their experiences and knowledge (Haraway 1988; Harding 2004). Looking critically at power can help us think through challenges. The *Feminist Manifest-No* is a project of feminist scholars in information sciences who have articulated this view (Cifor et al. 2019). For feminist approaches to research and ethics, gender "plays an important role in the reflection on how to decide" disagreements (franzke 2020). *Data feminism* shows how concepts from feminist theory might expose "the significant human efforts required by our automated systems" (D'Ignazio and Klein 2020). What this means for data science is that it is important to recognize that different people have different power in society, and data has the potential to harm people, denying their experience or denying their perspective or knowledge.

For example, consider the work that it takes to label large datasets through crowd platforms. This has been called *ghost work*, the often intentionally hidden human labor that powers mobile apps, websites, and artificial intelligence systems, keeping search results safe from adult content, tagging images that make machine vision look smart, and providing the human-in-loop work that makes content recommendations possible (Gray and Suri 2019). The invisibility of this work often results in lack of measurement of the work and lack of attention to the *people doing the work* (Star and Strauss 1999). If we do not recognize the invisible workers, then we may not record their data, and thus we may not know enough to be concerned about their working conditions or their wages and benefits. Angelika Strohmayer and colleagues provide a contemporary example of invisible work and invisible workers in their study with sex workers (Strohmayer et al. 2019), who are usually not considered in employment analyses or economic segmentations. If we cannot see the people— or if we choose not to collect and analyze their data—then how can we ask about their work and their well-being? Data justice, feminist, and Indigenist approaches also show how ethical practices, concerns, and commitments will necessarily vary across communities and in intersecting power relations of gender and "race, class, ability, sexuality, and immigrant status, and many more" (D'Ignazio and Klein 2020).

Science and technology studies teaches us that knowledge is shaped by the contexts where it is produced (MacKenzie and Wajcman 1999). Rather than a single truth, knowledge is constructed through negotiations that involve multiple ways of knowing the world. Consider community activists who want to map something about their neighborhood, like the prevalence of parks or grocery stores that sell fresh produce. Producing that kind of map might change the power the community has to push for improvements. For example, community activists collect sensor data to measure the air quality in their neighborhood to compare it to other neighborhoods. They may have a right to this data, but society may see scientists as having "better" or more "rigorous" data because of their status in society, even if the locals have more in-depth knowledge at the neighborhood level (Neff and Nafus 2016). People with particular medical conditions may want to share data and information about their condition to challenge established medical practice (Fiore-Gartland and Neff 2015). These imbalances of power in society around issues of science and technology mean that data science can have an enormous impact on people trying to make sense of their world and advocate for their communities.

Mixed Methods

Mixed methods is the term for social scientists using two or more methods on one project to take advantage of the strengths of both. For example, a user survey might be used along with interviews to allow people's particular experiences to give insights, background, or context for the survey findings. So far in this chapter, we showcased the benefits of the social science and design approaches and their potential usefulness to data science. In practice, combining multiple methods or extending them to accommodate each other might yield the most insight into people's behavior. In the rest of this chapter, we provide examples of approaches that combine data science techniques with other methods or extend them in ways that can account for and take advantage of the social context.

Taken together, multiple methodological approaches from domains outside data science have much to offer in how questions are asked, which contexts are accounted for, whose data is collected, and which results are reported. In the next section, we will look at how these methods can be combined with existing data science methods.

Combining Methods and Extending Data Science

Social science methods and approaches from design and critical studies can make data science more human-centered. We can also extend existing data science methods to take advantage of their computational power and account for how human behavior and activity are highly dependent on the context of where they happen. The goal is to provide computationally scalable results that can also account for the contextual aspects of human behavior.

Collecting Context-Rich Data

One of the simplest ways of incorporating more context into a data science pipeline is to collect *context-rich data*. By this we mean datasets that provide enough information about human activity, especially social interactions and cultures, to allow the data scientist to account for the various differences present in human behavior. This might vary across questions and datasets, but the main idea is to collect data that is sufficiently rich and does not exclude certain attributes of human behavior at the expense of other attributes.

Many data science analyses rely on social media data as a "trace" of human activity. The popularity of the social media platforms, the volume of activity on them, and the readiness of computationally readable data make social media an attractive source of data for data science projects. However, data scientists must be aware of several issues of context (recall the streetlight effect described in chapter 1). For example, collecting social media data by selecting for certain keywords or hashtags will limit the representativeness of the data. The sociologist Zeynep Tufekci calls selecting social media posts for their hashtag "selecting on the dependent variable," meaning that inclusion in the sample depends on the very variable being examined: "In hashtag datasets, a tweet is included because the user chose to use it, a clear act of self-selection. Self-selected samples often will not only have different overall characteristics than the general population, but they may also exhibit significantly different correlational tendencies which create thorny issues of confounding variables" (Tufekci 2014, 507). People also use social media conversationally (see chapter 5), referring to earlier posts for context and building on them as a background, instead of reusing keywords and hashtags in later posts (Palen and Anderson 2016). Collecting social media posts based on keywords or hashtags could miss this conversational context and likely ignore many relevant posts. In addition, social media posts are also part of conversations with others, through replies, reposts, and so on, which is also part of their conversational context.

Users of social media also have different demographic characteristics than the general population. Many social media platforms have particular features that are used differently in different countries (Miller et al. 2016). What works for looking at social media data in the US and Europe may not work in India, South America, or China.

One way to combat these limitations is to collect more conversational context around the social media data. You can collect keyword-related posts *and* other posts by the same users or a subset of users, based on some selection criterion. Instead of gathering just posts that use the hashtag for a major storm, the sample could include those people who are in the affected area (Kogan, Palen, and Anderson 2015) or who provide authoritative information about it (Bica et al. 2019; Li, Bahursettiwar, and Kogan 2021). Another way to do this is to add interactions among users who produce keyword-relevant posts and the people who engage with them by, for example, sharing their posts, replying to them, or mentioning them. These resulting datasets are much richer in social context and capture people's interactions. Such datasets also let you explore the variety of ways that people communicate and the diversity of people with whom they communicate.

Social data is one meaningful resource and means of collecting context-rich data. However, data scientists who work in other settings such as public services or clinical settings also look at historical narratives about a system, practice, or process to understand context. For instance, researchers examining how algorithms are used in child welfare services (Saxena et al. 2020) are using historical caseworker narratives to understand questions—for example, Which strategies for deciding when to move a foster child between homes work best on the ground when confronted with scarce resources? Similarly, researchers have been analyzing patient narratives (Demner-Fushman, Chapman, and McDonald 2009) that are typed up and stored in electronic health records. An analysis of narratives can provide effective strategies for serving different kinds of patients and perhaps even provide insight into strategies for dealing with rare conditions or symptoms that individual doctors may not have seen but that can be compiled from historical, organizational records.

Keep in mind that contextual information about the dataset is often crucial for understanding the meaning of the data. Data collected for a specific purpose, with a specific question in mind, might not be well suited for answering different, even related questions, because of what has been measured and how (see also chapter 3). An excellent example is a series of papers done at Microsoft Research around predicting social behavior from social media like Facebook or Twitter (DeChoudhury et al. 2013; Murnane and Counts 2014). The group was interested in predicting many kinds of behavior from social media: for example, depression, anxiety, and smoking cessation. However, one of the main lessons learned over a period of six years is that the same algorithms and features that are used to predict depression on Twitter don't work to predict even something closely related like bipolar disorder or anxiety. Thus, they worked with domain experts like mental health clinicians to develop questionnaires to help them more accurately analyze these very different outcomes using public social media data.

Incorporating Context into Data Science Methods

In data science we often expand our existing methods to take advantage of new types of data. Capturing aspects of social behavior that are outside of the data itself, like shifts over time or changes in the underlying definitions (e.g., what is defined as "sexual assault" in the United States has changed in the last ten years in different criminal justice systems), can take advantage of the presence of more structure in the datasets and thus enable the development of more powerful models.

Natural language processing (NLP) may be a good example of where this approach can be applied. NLP is a subfield of linguistics, computer science, and machine learning that is concerned with enabling computer systems to process natural language—how people naturally speak. In practice, that often means building algorithmic tools that can process—that is, translate, categorize, or reveal the semantic structure of—large amounts of text or voice data. Textual data may reflect a conversation between people, like social media posts. It may reflect relationships among things, such as Wikipedia pages that link one article to another. Textual data may reflect sequences over time, such as medical records. If collected over time, textual data is likely to be dynamic and changing: the conversations on social media, for example, can shift rather dramatically over the course of a month.

Considering this, human-centered approaches to NLP should account for and take advantage of this context. One approach to incorporate context is to use changes and differences in the data to the advantage of your study. For example, if a dataset of tweets changes before and after a major event—like a storm—then we can use this knowledge to develop more nuanced models that can detect these changes. This approach has been successfully applied to *topic models*—statistical models for discovering abstract topics in collections of documents. Topic models can algorithmically summarize large quantities of text by grouping or clustering them around particular topics. Topic modeling is based on the frequencies of word co-occurrences in documents. Researchers have extended topic modeling to also detect temporal changes (Paul 2012). This approach allows you to detect both topics at a particular point in time and changes in topics over time. This incorporates temporal context into the data and produces a more nuanced, powerful model.

Other methods inherently incorporate context because of the fundamental assumptions they make about relationships in data. One such method is *network science*, which applies graph theory approaches to sets of relationships, including interactions between groups of people, known as social networks, or relationships between things such as telecommunication links or computer networks (Wasserman and Faust 1994; Monge and Contractor 2003; Newman 2018). Network science represents relationships among distinct actors, known as "nodes," who are connected to one another through links or social ties/relationships (also called "edges"). The set of these relationships are shown as a "network," which includes both nodes and edges. The data structure in network science presumes relational properties among the data points. Nodes are not independent; they are interconnected through relationships in the network. This view of the world maps well to human and social media data, where people and their activities are not isolated but rather are interconnected through a complex web of relationships. The same types of analyses can be applied to documents that link or refer to one another, like academic or scientific literature or legal rulings. Network relationships are a source of social context in human-centered data science.

This makes network science a context-rich, potentially human-centered data science method. In addition to large-scale results of network analyses (network structure, community detection, centrality), network science also preserves some of the context of the social interaction because of its use of relationships in data. For example, a network analysis would represent your Facebook friends as a map of nodes and edges. Because network analysis retains the relationships between data—you are connected to your friends—it preserves local social context in large-scale results. Network science also allows us to observe and model how small-scale local behaviors—tagging your friends, using a specific hashtag— "add up" to the larger social dynamics, such as reflecting a trend or contributing to an event. In network science, the more context gets captured in the network, the more precise results the models produce for the large-scale questions of network structure and dynamics.

As we mentioned earlier, Shion and Michael examined another property of networks when they examined how a type of employee experience spread from one employee to another, transmitted through the employees' networked relationships with one another (Muller, Shami et al. 2016). Employees (nodes) were connected to one another through their formal work roles (organizational, or org, chart) or their friendships (social network ties). Both org-chart and social ties became edges in the network. By analyzing these different types of edges, Michael and Shion could see which types of edges, or types of relationships, were more important in the spread of the employees' experiences (see also Mitra et al. 2017).

Mixing Methods to Incorporate Context into Data Science Methods

Recent innovative advances are combining various methods, both from social science as well as existing data science methods, to bring much-needed context into data science methods.

Grounded theory + topic modeling: Multiple approaches to data science methods share the objective of dealing with large amounts of text data that have a high chance of built-in social context. Here, we consider two different methods that both make strong claims to be "data-driven," beginning from a method-to-method comparison that Michael and Shion made (Muller, Guha et al. 2016). Grounded theory relies on close reading, constant comparison (of data with data, and of data with theory), and theoretical sampling (using what you know to find the weakest point in your theory and then collecting data—that is, "sampling"—to test that weakest point). Researchers use methods developed from grounded theory and related techniques to infer which underlying themes describe and represent the main contexts within text data (Charmaz 2006; Corbin and Strauss 2014; Muller 2014). By contrast, topic modeling is a computational method in the NLP family of unsupervised machine learning models that looks for statistically regular patterns of words within a large corpus of text data to find statistically valid "topics" of interest (Blei, Ng, and Jordan 2003).

While the topics are purely derived from statistical regularities in the text, data scientists often have little idea what they might actually mean. This presents an interesting opportunity for data scientists seeking to add context to their work to use methods and tools of analysis adapted from grounded theory on part of their data. The catch is that grounded theory takes time and is hard to scale, yet it describes the rich human context that we so deeply desire in data science. Recent methodological advances combine grounded theory methods and topic modeling, with the rationale that "small" amounts of deeper analysis on text data can be used to "seed," scale up, and add context to topic modeling in useful and interesting ways that move beyond statistical text patterns. For example, researchers used methods from grounded theory and topic modeling in parallel to understand which contexts were useful in predicting when someone might deactivate or delete their Facebook account (Baumer et al. 2017). Researchers found that the themes that arise from the closer textual analysis did not map directly to topics that emerge from topic modeling. Used with topic modeling, grounded theory methods can bring context to inform, iterate, and change topics derived from topic modeling to scale up through thousands of data points.

Network science + ethnography: Another interesting way to combine multiple methods to add context to data science methods is to integrate network science models and grounded theory methods. Earlier in this chapter, we introduced network science as a data science method that helps researchers infer relationships, connections, and power between people and entities. For example, researchers seeking to develop models of scientific collaboration have analyzed large-scale co-authorship, citation, and co-citation networks using network science models to identify specific successful people who have been subsequent participants in ethnographic investigations of their work and their research groups (Velden, Haque, and Lagoze 2010). The "network ethnography" method uses social network methods to justify case selection for ethnographic or qualitative data collection (Howard 2002).

Say you do an analysis of groups of scientists to understand how they introduce and adopt new ideas. You could use the network maps of their collaborations to sample scientists for interviews or observational study. For example, Marina has conducted a network analysis of how crisis mappers collaborated and worked together on mapping Haiti in the wake of the 2010 earthquake on the OpenStreetMap platform. The structure of the network she derived from the mapping activity allowed her to select mappers who worked together with many others and those who did not. She then conducted in-depth interviews with a subset of these mappers to further understand and contextualize their collaborative practices (Kogan et al. 2016). This is an example of a top-down approach: starting from a large-scale network structure and then diving into more fine-grained experiences and meaning. Both top-down as well as bottom-up holistic investigations of complex, social systems (such as scientific collaborations) are very good examples of how methods can be mixed to achieve success in solving data science problems.

Visualization and Reflection

Sometimes it is not possible or useful to build machine learning models for particular types of research. Cecilia and her students built a visualization tool to help qualitative researchers streamline their labeling of short pieces of social media text (Chen et al. 2018). Although they planned to use machine learning, in initial studies they discovered that researchers were more interested in exploring inconsistencies and ambiguities than coming up with the definitive "ground truth" label for each piece of data. These ambiguities often point to potentially interesting nuances in the data. Thus, one of the prime benefits of this visualization tool ended up being that it offered the opportunity for its users to reflect on the data, collaboratively identifying points of interest, unanticipated complexities, and intriguing nuances that might lead to insights, rather than improving labeling efficiency on an arbitrarily defined "truth."

This ability to reflect is often fostered by visualization. One advantage of developing a visualization tool informed by human-centered methods is the ability to add in features that support reflection and thus greatly enhance the power of the visualization. For example, sarcasm is notoriously difficult to identify using automated means. In Aeonium, the visualization tool that Cecilia and her students built, users can label data with both a code and an ambiguity label. Participants who evaluated this tool provided open-ended feedback to identify the most common features leading to ambiguity. Potential sarcasm ranked very highly among them. Visualization tools can color-code or otherwise identify ambiguous data and provide means for users to add their own open-ended feedback.

This suggests that identification of dimensions of ambiguity may help improve feature engineering and model tuning in machine learning tools. In other words, the reflection facilitated by this visualization tool both pointed out interesting areas for future social science work and presented opportunities for algorithm developers to improve the effectiveness and accuracy of their software. At the eScience Institute at the University of Washington, two of the authors (Cecilia and Gina) have seen examples of what we call the "virtuous cycle" of data science across multiple fields. Cecilia's previous work at the intersection of human-centered data science and astrophysics both advanced physics and inspired the development of new data science algorithms. Her experience of being in two areas is not unusual. Additionally, the type of reflection facilitated by visualization, particularly collaborative visualization where people with training in very different fields come together over a single image, can be very powerful and lead to new knowledge in multiple areas.

Data Science Ethnography

Ethnographic methods are useful for studying data science in action, focusing on how people "work with large and complex datasets, and the institutions, programs, and communities" that support data science (Aragon and Poon 2007, 2011; Aragon et al. 2016;

Aragon et al. 2009; Brooks et al. 2013; Chen et al. 2018; Poon et al. 2008; Tanweer, Fiore-Gartland, and Aragon 2016). Data science ethnography brings together tools of ethnography and data science itself. At the University of Washington eScience Institute (where two of the authors have worked), this meant training ethnographers to help "critique and contribute" to building a large data science project, looking in-depth at how people use and learn data science methods and how different communities make sense of and value data (Kuksenok et al. 2017; Neff et al. 2017; Rokem et al. 2015; Tanweer 2018). Wikipedia has used ethnography and data science together to help design a better way to suggest citations to Wikipedians (Ford 2014). In studying computer vision and self-tracking, anthropologists suggest that much is to be learned from focusing on all the steps that go into a data science project: "the data preparation, the coding, the learning and application of all those rules, the repeated experimentation and testing and the debates over what the algorithms actually do and then what they do in someone else's hands and with someone else's intentions" (Thomas, Nafus, and Sherman 2018). One way to think of data science ethnography is to view data science as a part of culture and as producing a culture that influences how people move through the world. Another is as a way to reflect on data science's methods and practices and learn from that reflection. Yet another is to see how the study of people in their context can shape data science projects.

Iterating among Levels of Analysis

In figure 6.3 we see a type of iteration between levels of analysis. At the heart of this human-centered technique is the idea of shifting between micro-level (small-scale, local) behaviors, like your Facebook interactions, and the macro-level (large-scale) dynamics, like the overall prevalence of misinformation on Facebook. Shifting between the individual or small-group behaviors (micro) and large-scale social dynamics (macro) allows these levels of analysis to inform and enrich each other.

One approach is to use large-scale data science analyses to broadly understand and model a specific phenomenon. These tools are often ill-suited for more nuanced, in-depth investigations, especially those focused on understanding why people behave in certain ways. To compensate for these limitations, we can use data science methods to scope a subset of data for a deeper analysis—for example, by finding an interesting and informative subset of people or behaviors that will lend more insight into the phenomenon when analyzed in detail. This "deep dive" into a specific subset of the data is often done using qualitative methods, as these methods are better suited for understanding particular perspectives and experiences of people, as well as for answering the questions around *why* people behave the way they do.

An alternative approach would be to start with a deep, small-scale analysis that provides detailed insight into specific phenomena. Again, this is often better done using qualitative methods. One of us (Michael) was suspicious of published claims (e.g., Gray et al. 1993, 1995) that telephone operators performed only rote, thoughtless tasks repeatedly. Working with telephone operators as co-analysts, Michael and colleagues performed a small-scale, in-depth qualitative participatory analysis that suggested that telephone operators performed diverse forms of knowledge work. Next, the team showed that operators performed knowledge work in 53 percent of a random sample of customer calls (Muller

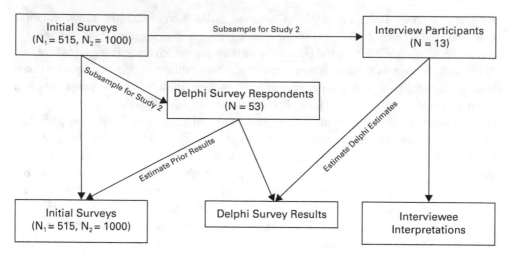

Figure 6.3
Combining small-scale qualitative analyses with large-scale quantitative analyses. The analyst can extract a useful concept from each type of analysis and use that concept to reshape or "tune" the next, alternating type of analysis.

et al. 1995). The project contributed to analyses of *invisible work* mentioned earlier in this chapter, which may have been underrecognized because of the gender-marked nature of the operators' job (Muller 1999). The small-scale qualitative analysis provided information about operators' work for the large-scale quantitative analysis.

In-depth qualitative analyses can serve as an important exploratory stage when the scientist does not know enough about the domain to decide what questions might be important to pursue quantitatively (Muller 2014). This initial in-depth analysis would then allow the scientist to better understand which types of questions might be interesting and thus generate more nuanced hypotheses that can be tested with large-scale methods. Here the scientists might use data science techniques or other quantitative methods like survey research to arrive at more generalizable results.

Great strides have been made recently in thinking about data science solutions to problems that can dynamically iterate between different levels of analysis. For instance, researchers have developed methods that are flexible enough to iterate between units of analysis when required to do so (Velden, Haque, and Lagoze 2010). In a project studying scientific collaboration between large groups of chemists, these researchers developed what they call a "mesoscopic" analysis to oscillate between different levels of analysis when necessary. For instance, they used social network analysis of co-authored scientific publications to pinpoint individual scientists that they would need to observe or interview, which meant the latter would become the level of analysis. The reverse was also conducted to move from the individual scientist to the unit of the scientific publication in a co-authorship network.

Conclusion

Looking at key traditions of social science, design studies, and critical theories shows us how other fields deal with some of the challenges of studying human behavior.

First, we traced social science methodological approaches in qualitative, quantitative, and design methods. Qualitative methods include ethnography and observations. We also looked at quantitative and computational social sciences approaches, as well as transforming methods and techniques from design and critical studies. We concluded with a section on some of the innovative ways of combining different methods to facilitate better data science.

Our goal is to show how these approaches can inform new ways of leveraging computational tools with the abilities people have for interpreting, building and transforming the social world we live in.

Recommended Reading

Babbie, Earl R. 2017. *The Basics of Social Research.* 7th ed. Boston: Cengage Learning. Good general introduction to qualitative and quantitative social research methods.

Institute for the Future and Omidyar Network Tech and Society Solutions Lab. 2018. "Ethical OS." https://ethicalos.org. A set of exercises and checklists for thinking through ethical implications of new technologies, targeted to industry professionals.

7

Collaborations across and beyond Data Science

Who does data science work? In practice, data science is often accomplished in teams—collaborative, cross-disciplinary teams involving both people who are experts in a field and those who have training and expertise in other areas (or all of the above). In this chapter, we look at the specific roles that people play in data science teams, and we put those teams into a broader context of organizations and social values. Our discussion begins with the "sociotechnical" interactions of groups of people with technology. Data science teams often involve people with different types of knowledge, and we therefore consider which disciplines contribute to data science work. Then our focus expands to the collaboration of people with artificial intelligence (AI) in data science, beginning with simple, pairwise interactions and moving on to six patterns of human-AI collaboration.

We also examine the organizations where teams work and the larger-scale communities in which data science applications may have impact for good or ill. Even when the data does not specifically concern humans, sociotechnical factors matter, and we illustrate this with an example from science. We conclude with a statement of the stakes and ethics of these collaborative relationships.

Working in Teams

Let's begin by considering the teams of people who make a data science system. Data science has always been a "team sport" (Aragon and Poon 2007, 2011; Aragon, Poon et al. 2008; Aragon, Poon, and Silva 2009; Aragon and Williams 2011; Finholt and Olson 1997; Ohana n.d.; Olson, Zimmerman, and Bos 2008; Poon et al. 2008; Stein et al. 2017; Wuchty, Jones, and Uzzi 2007; Wang et al. 2019; Wulf 1993). Years of evidence have shown that data science problems require multiple skill sets (Aragon and Aragon 2007; Aragon, Bailey et al. 2008; Aragon et al. 2009; Kim et al. 2016; Matsudaira 2015).

People from different backgrounds and with different roles often participate in data science activities and often participate together on a team (Aragon, Bailey et al. 2008; Borgman, Wallis, and Mayernik 2012; Dossick et al. 2015; Dossick and Neff 2011; Dossick, Neff, and Homayouni 2009; Dossick, Osburn, and Neff 2019; Passi and Jackson 2017; Vertesi 2015; Zhang Muller, and Wang 2020). You probably think of yourself in the role of a data scientist. Here we want to explore other roles on the data science team and their work practices, collaboration patterns, and tools.

- Think about a time (present or past) when you were working on a project with a team of other people. What were the dynamics of your team? How did you share your knowledge? How did your team members share their knowledge with you?

- Have you worked on a project team with people who have backgrounds different from your own? Compared with your answers to the previous question, did you experience different dynamics or different forms of knowledge exchange on these multiple-background teams? How did you combine your different types of knowledge to solve the team's problems?

Working with Other Disciplines

One of the major advantages of working in a team is the combination of diverse types of knowledge. For a data scientist focused on algorithmic issues, it can be difficult to develop a deep understanding of any particular situation or context of the project data. Domain experts—people who are deeply knowledgeable about a specific type of problem or field—may ask questions that have more relevance. For example, when one of the authors (Cecilia) joined an astrophysics collaboration as a data scientist, she learned that astronomers' most frequently asked question during a night of telescope observation was "When is sunrise?" This surprising (to her) observation and others led to the creation of a chatbot that increased the amount of successful data collection by a factor of four (Poon et al. 2008).

Domain experts can ask more pointed, more interesting, and more novel questions about the domain of the data science project. They are more likely to know what type of work has been done in this domain, what metrics are available for measuring certain aspects of behavioral or nonbehavioral outcomes, and what datasets or data collection methods would facilitate more in-depth analyses. Domain experts are also likely to have a "feel for the data," with an ability to sense when data patterns "don't look right" (Muller, Lange et al. 2019; Wang et al. 2019). They may also be able to recommend different ways of classifying data and different combinations of data for novel engineered features. Thus, collaborating with a domain expert is an excellent way to gain deeper understanding of the domain, the potential questions, and datasets, which ultimately produces a more precise, in-depth analysis of the behavior. Nevertheless, data scientists can also possess deep training in a domain or specialty providing insights that span disciplines.

On the other hand, collaboration with data and computer scientists may help domain experts think in more computational, data-driven ways. Consider an example from one of our universities where data scientists joined the oceanography labs. Having people work on coding projects in the lab of a large oceanography team changed how oceanographers asked research questions. Even though they did not always use the software provided by computer scientists, oceanographers shifted to asking questions that were more informed by the computational perspective (Kuksenok 2016). Vertesi (2015) reported similar knowledge-based insights from working with teams that collaboratively programmed the Mars rovers, as did one of the authors (Cecilia) in her work with astrophysicists (Aragon and Aragon 2007).

While there are many advantages in working with multidisciplinary teams, there are also challenges. An excellent example comes from a study in the field of ecoacoustics—a discipline that "draws together computer scientists and ecologists to achieve an understanding of ecosystems and wildlife using acoustic recordings of the environment" (Vella et al. 2020, 1). The study found that when computer scientists joined a large ecology lab, all the scientists encountered "different temporal rhythms, relationships to data, and data-driven questions." Some of the difficulties of working together had to do with the technical aspects, such as what was considered data and what kind of questions were appropriate (Mao et al., 2020; Pine and Liboiron, 2015), but others had to do with more social factors, such as how fast different disciplines move through their projects.

- Consider a project you are excited about that may require working with a person from a different discipline. Which aspects of this collaboration might be easy? What difficulties do you anticipate?

Pi-shaped People

The phrase "T-shaped people" has been used for a long time in education as well as marketing, business, and technology to refer to individuals with deep training in one specific area—the vertical line of the T—but also a broad but necessarily shallower background in multiple related areas—the horizontal line of the T. People use the phrase "pi-shaped" (referring to the Greek letter π) to describe people with deep expertise in more than one area—the two deep vertical lines of the "π" symbol. Such individuals are often extremely helpful members of data science collaborations, in part because of their ability to translate vocabulary and insights from one domain to another.

Additionally, people with expertise in multiple areas can often serve as interpreters between experts in various domains (Williams and Begg, 1993). They can speak the language of multiple domains and thus act as translators or bridges between groups that may otherwise come into conflict.

Disciplinary Bias

Tony Becher studied academic disciplinary cultures in the 1980s and described "disciplinary bias" as the unwillingness to appreciate others' domains (Becher 1989). It's not a surprising occurrence. Most people train for and develop expertise in an area that they believe is important. Over time, people naturally tend to develop a belief in the ascendance of their own field over others. As a result, they may consider people who are unfamiliar with the terminology of their field as somehow less intelligent or skillful. Certain words carry emotional as well as gateway connotations: how you refer to a particular topic reveals to others that you are a member of the same in-group. This is particularly true in matters of expertise. For example, many people are familiar with the term "machine learning," so using this term is relatively neutral within this group of people. But if you mention "support vector machines," you are signaling to others that you have at least an intermediate level of knowledge in the topic. Bring up "contrastive divergence," and you are signaling a particular type of knowledge and higher-level expertise.

For many experts, their expertise is hard-won, often over years, and in many cases it may be core to their identity or at least their income potential. As a result, showcasing this expertise is vital to their self-presentation among both members of their field and those outside it. Thus, when others introduce methods that are outside these experts' areas, particularly by using words that they don't understand, it may lead to dismissal of the new ideas or even of the people who present them. This tendency, like other forms of bias, may very well be unconscious. Further, if people who consider themselves experts are introduced to terminology they have never heard of, they may feel diminished or slighted, even if they are not consciously aware of it.

For this reason, we recommend that you pay attention to your language when you join a new data science team. Make sure to explain complex terminology using simpler building blocks. One way to do this is to introduce the term and then immediately define it. For example, when one of the authors (Cecilia) described the use of ethnography to a team of physicists, she promptly explained the term and, more important, explained exactly why it would be helpful to the physicists in their work.

This type of awareness of more subtle human interactions, which may be missed in traditional data science training, will serve you well in a data science career.

- Which words do you use to showcase your own expertise? Think back to a time when you were introduced to another group and members of this group used words that you didn't understand. How did you feel?

- Have you ever felt unintelligent around a group of people? What led to this feeling, and what could have alleviated it?

- Have you ever seen someone's eyes glaze over as you explained a concept? Can you think of ways to avoid this?

Working with Data Science Teams

In this section, we describe the most common roles assembled on data science teams. We consider the communication patterns of these different roles and the tools they use for the team's collaboration work as well as the technical work of data science.

A survey of more than 180 data scientists at IBM asked people to self-identify into one of these five roles (based on an earlier Kaggle survey of data science workers, by Hayes 2018):

- **Researchers/scientists** ask and answer abstract questions. To do this, they often use data science tools and software packages.

- **Engineers/analyst-programmers** carry out more detailed programming work. They often perform some of the "grunt work" of data science, cleaning datasets, engineering features, and so on.

- **Domain experts** (if available) provide knowledge of the target application domain of the system being built. A domain expert may be a medical professional, a financial analyst, a bank officer, a community leader, or a member of the workforce that will actually use the data science application that the team is making.

- **Communicators** design and sometimes deliver communications across team boundaries. In practice, their work can involve both receiving materials from others and creating and providing reports to people outside the team.

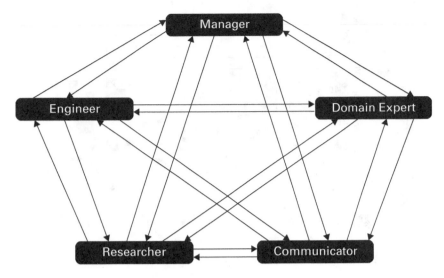

Figure 7.1
Patterns of collaboration among members of data science teams. Teams in other studies include additional roles, such as statistician and data scientist. See Muller et al. (2019) for a more extended list.

- **Managers/executives** represent the needs of the organization during the work of the team, and they also coordinate the work within the team.

It is important to remember that this case study was limited to employees of one company, and that often, members of data science teams do not perform a single role. All data scientists must act as communicators to share their results with others (as discussed in chapter 8). However, this study is nevertheless valuable for its analysis of collaboration patterns within a large set of data science teams. The study found at least 87 percent of IBM employees in all roles reported that they collaborated with others on data science projects (Zhang, Muller, and Wang 2020). Figure 7.1 shows a summary of collaborative relationships among the roles. The edges connecting each role to another role indicate the average frequency of the collaboration, as reported by each role. We can see that each of the five roles collaborated with each other. Interestingly, many teams included a domain expert as a member of the team.

During the intensely technical phases, much of the collaboration involved "implementer" roles such as engineer and analyst-programmer, and the role of "communicator" mostly disappeared (see figure 7.2). Communicators were more involved in collaborations during the earliest phase of the work and then again at the end of the cycle. Unsurprisingly, communicators' collaborations were strongest during the phase of reporting outcomes to people outside of the team. They also participated in the preceding phase of evaluating outcomes, perhaps in preparation for their work reporting to outsiders. Finally, Zhang, Muller, and Wang (2020) asked about the tools that the teams used at different phases of the data science cycle (see figure 7.3). Much of the work was done in chat spaces such as Slack rather than in face-to-face or online meetings—similar to the findings from a series of earlier studies in a different workplace (Aragon, Poon et al. 2008; Aragon et al. 2009; Aragon and Williams 2011). The only phases in which there was greater use of synchronous meetings was during work on the measurement plan and the reporting-out of the project outcomes. Despite the team structure, little of the work was done through real-time interaction. File sharing

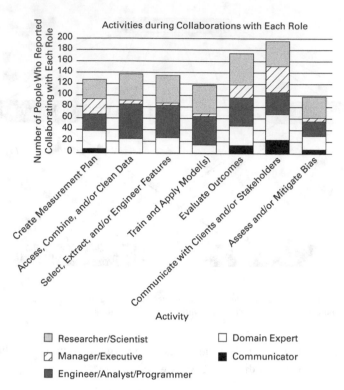

Figure 7.2
Patterns of collaborations across the data science cycle.

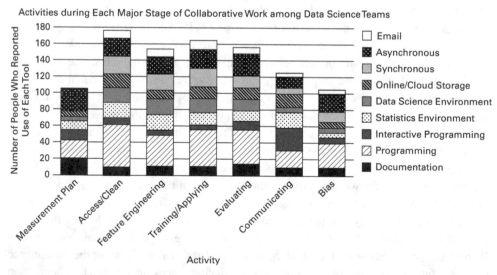

Figure 7.3
Tools used during different phases of the data science cycle.

occurred frequently in the early stages, dropped off to near-zero during the intensely technical phases, and reappeared during evaluation of the outcomes. The technical phases were
spent primarily in coding environments (Zhang, Muller, and Wang 2020).

When data science teams did edit documents, it was primarily during the phases of measurement plan, evaluating outcome, and communicating outcomes. These were the three
phases when there was relatively strong collaboration with the communicator role (see figure 7.3). This suggests that the document editing was done primarily by the communicators.

- What types of records do you keep in your projects? Do you ever forget some details
  of a project and then need to spend time *re-creating* those details? Have you ever had
  to spend a lot of time helping someone else understand what you did?

- Have you ever received data or code from someone else and then had to ask them to
  explain what they did? Did they provide documentation for you? What did that documentation tell you? What would you have liked to have known about their code or
  data that was not included in their documentation?

- How do you imagine that other people would use your documentation while exploring or interpreting your data and code? What would help them?

- How would you document your work with data and/or code? In what medium would
  you create and keep your records? Is there a way you could attach your documentation to the data or the code? What would you need to do to bind code, data, and documents together?

- If you were to manage a data science team, how would you incentivize your team
  members to reflect on their process and keep more documentation?

Understanding Patterns of Human-AI Collaborations

We now shift to looking at how humans might collaborate with an AI, such as an AutoDS
application (see chapter 4 for more on AutoDS).

Biles (2020) proposed four core patterns of human-AI work to create ideas or projects.
We supplement these patterns of *making* or *planning* with a preceding step of *sensing* (to
obtain information that can be the basis of making or planning) and a culminating step of
doing or *executing* (to take the action that was envisioned while making or planning).

- Sensing (autonomously)
  - Human delegates the task of collecting data to the AI system. In many cases, the
    AI system operates autonomously to collect the data and then provides the data-
    set to the human.

- Making or Planning (Biles 2002)
  - AI system generates alternatives, and the human chooses among them.
  - AI system generates alternatives, and the human chooses one of those alterna-
    tives and modifies it.
  - Human generates alternatives, and the AI system selects and modifies one of
    them.
  - Human and AI system iterate collaboratively to create and modify a solution.

- Executing/Doing (autonomously)
    - Human delegates an action to be performed by the AI system, and the AI system performs it. In many cases, the AI system operates autonomously and reports completion to the human.

These core patterns help us think about how human-AI collaborations are initiated. We begin with two unidirectional patterns:

1. The AI "makes the first move" to present an idea to the human. Depending on the domain or application, this could be that an insurance algorithm determines whether an insurance claim will be paid, or a "doctor's assistant" AI reads a patient's electronic health record and then recommends medical tests (Dilsizian and Siegel 2013). The important aspect is that the AI initiates and the human takes note of what the AI has communicated, and the collaboration stops at that point.

2. The human initiates, and the AI responds. Perhaps the human asks for a recommendation of a movie to watch (Hallinan and Striphas 2016). The AI answers, and the collaboration stops at that point.

Two additional patterns involve a greater degree of interaction among human and AI:

3. In creative domains, the human generates an idea, and the AI elaborates on it. An example is the Bach Doodle (see also chapter 4), in which a human provides a musical theme of a few notes and the AI produces a four-part chorale based on that theme, in the style of Johann Sebastian Bach (Huang et al. 2019).

4. In the most interesting pattern, perhaps, because it is more like a conversation, human and AI take turns to work on a concept or an artifact, such as the interactive drawing application DuetDraw (Oh et al. 2018). Other examples include human and AI taking turns writing a short essay or a slogan (Clark et al. 2018), or human and machine engaging in a chat to achieve a scientific objective such as creating an astronomical observing schedule (Poon et al. 2008).

Biles's patterns were primarily about making and/or planning. However, people and AI systems often need information before they can do this making or planning. Therefore, we propose an antecedent pattern:

5. The AI gathers data in response to a human specification about what kinds of data is needed. Unlike Biles's patterns, there may be a longer time interval between the human's instruction and the AI's return of data, such as in a programmed Mars rover (Vertesi 2015) or in collecting data during an emergency. In this case, the AI could conduct substantial parts of its action without direct human supervision (Winner 1978).

Finally, people and AIs also need to act:

6. The human states an intention, and the AI carries out that intention without further human guidance. Sometimes, this happens immediately, as with commercial aircraft in which the human pilot indicates their intention and the plane's onboard AI carries out the action with micro-adjustments that become necessary as flight conditions change (Miller 2014). In other cases, the human gives a command and the AI carries

it out until the action is completed, such as the delivery of emergency supplies (Chowdhury et al. 2017). This could be a matter of hours, such as when a drone is told to return to base (Nguyen et al. 2018). There are also many well-known examples where insufficient attention to the interface between human and AI led to aviation, ship, or other disasters (Degani 2004). It's important to carefully imagine what the AI might end up doing and program-in stopping conditions or allow for human intervention. As the Universal Paperclips game mentioned in chapter 1 illustrates, unexpected consequences are some of the many common problems with this pattern of human-AI collaboration.

These six patterns help us think about how a data science system could be used.

From a human-centered perspective, an understanding of use should inform design of both the user experience and the technology that shapes it. It is also important to consider where these patterns could go wrong, or what unintended consequences could occur. These are merely starting points, because good human-centered design requires working directly with the humans who are involved. Minimally, this means working directly with the intended or imagined users. As we will explain later in this chapter, human-centeredness can also include diverse other roles that have a stake in the design and/or use of the technology.

Ask yourself:

- For the data science system that you will build, what human-AI pattern or patterns would be most appropriate? Do you know how your future users are currently doing their work? Can you choose a human-AI pattern that would fit into their current work practices?

- What additional or new human-AI patterns can you imagine? What are their advantages and disadvantages?

Working in Organizations

In most cases, a data science system or service is implemented by or for an organization to achieve its organizational goals. Organizations may conceive of their goals as involving service to employees, customers, and the public, or as maximizing organizational metrics such as cost containment or risk avoidance. For large-scale data science systems, there are often many people involved, potentially in multiple roles. Different people and different roles may have convergent or divergent needs and interests. The *Value Sensitive Design (VSD)* approach offers conceptual tools that can help us think about the interests of different people involved. VSD is concerned in part with human relationships to designs, technologies, systems, and the policies that motivate them (Friedman and Hendry 2019).

A major concept in VSD is that of the stakeholder. A *direct* stakeholder is typically someone who comes into direct contact with a design or a system. In conventional software engineering language, this would be a user. It can also include people who create and build the system, such as engineers, developers, designers, and so on. VSD goes beyond most other analytic schemes in its concept of *indirect* stakeholders—the people who have an interest or a concern for the system or who experience system actions indirectly.

For a commercial customer-care support system, the direct stakeholder may be a customer-care service representative who answers telephone calls from a customer. The customer does not interact directly with the system but is affected by the service representative's experience. In this way, the customer becomes an indirect stakeholder. Other indirect stakeholders may include managers of the company or members of the customer's family. Suppose that the company provides medical insurance. In this case, the customer's family members and their physicians may also be indirect stakeholders.

Insurance companies have complicated goals: they provide payments to customers, but they also enforce various cost-control protocols that can prevent or deny payments to customers. These benefits and constraints are experienced differently by different stakeholders. We can use this framework to think about different people involved with greater detail and greater insight.

From a customer's perspective (indirect stakeholder), the service representative (direct stakeholder) may be perceived as exercising power over the customer's medical benefits. However, from the service representative's perspective, their scope of action may be entirely limited by the support software (the data science system). If neither the customer nor the service representative has any power, then who *does* exercise power in this configuration? One answer might be that the developer or designer of the data science system has power (direct stakeholders). However, development teams often "code to the spec" (specification), and they may experience powerlessness in relation to the person who wrote the specification for their system (yet another indirect stakeholder). We can continue to follow the chain of responsibility up through the management of the insurance company until we find the person who determined the policy about customer benefits versus cost controls.

We can also consider the COMPAS system that US judges (direct stakeholders) use in some states to decide how long a jail sentence should be imposed on a person who has been convicted of a crime (Eubanks 2018; Shin and Park 2019). The defendant (indirect stakeholder) has very little power to influence the judge's decision. What is less visible are the constraints on the judge, who will be criticized for deciding on a very brief sentence if the defendant commits a new crime when released, or for giving a very long sentence while other people advocate for kindness and returning prisoners to their families after shorter incarcerations. Judges are supposed to consider the COMPAS algorithm recommendations, along with other considerations such as sentencing guidelines and special circumstances. But they may face harsh criticism and even disciplinary action if they deviate from the algorithmic recommendations. As in the previous example, we may have to search at length for the indirect stakeholders who decided that there should be an algorithm that determines sentencing, and another set of indirect stakeholders who created the specifications for that algorithm.

Nonetheless, organizations want and need data science systems. We have so far explored how some organizational goals may require further analysis. While readers and the authors of this book are engaged in data science, we are all simultaneously users of other people's data science systems as well.

- Consider a data science system that you have built. Can you develop a *dual awareness* of that system—once as the creator and once as an imagined user? The roles of creator and user are both direct stakeholder roles. Can you imagine the experience of an indirect stakeholder of the system?

- In your stakeholder roles, what questions would you want to ask? What kinds of answers would satisfy you? How does that dual awareness change the ways that you would build a data science system?

Working with Communities

The word "community" may appear simple, but in fact, it means many different things to many different people. Data science is often concerned with working with different communities in different contexts. We outline here broad principles about how to work with communities that have populations that can be characterized as vulnerable in some aspect. Some examples are foster children in the child welfare system, incarcerated offenders in the criminal justice system, or gender minorities in the Global South.

Increasingly, data science is positioning itself as a field that not only can develop recommendation systems, dynamic advertisements, and product placements for large technology organizations, but also can be a force for social good (Tanweer 2018). However, it takes work and care to do data science well, and successful projects "allow stakeholders to tinker with the data at multiple stages of the process, as a mode of collaborative and public experimentation" (Zegura, DiSalvo, and Meng 2018, 7). Unfortunately, some data scientists persist in their belief that with enough data and computational elbow grease, many community problems can be solved (Tanweer 2018).

Data scientists must first ask themselves how their participation might responsibly support communities. For instance, what happens if a data science solution to a community's specific issues backfires and worsens the situation? MIT graduate students answered a hackathon challenge to "fix" the challenging problem of school bus schedules in Boston. Their solution was efficient, elegant, and data-driven. But implementing it unleashed a tangled mess of racialized housing patterns and city and school board politics, leading to angry parents, a political crisis for the city's mayor, and, at least in part, the resignation of the superintendent of Boston's schools (Scharfenberg 2018). On the other hand, conducting an action research method by engaging with communities with the intent to be participatory, empowering the communities and improving their lives, can be considered an ethical objective (Hayes 2011).

However, working with communities takes time and effort. Building a partnership means that "the partners aim to achieve something they could not do alone, by pooling skills and other resources. To do this they need a shared vision of their goals, and a way of working together" that realizes this ambition (Wilcox 2004, 1). There may be legal and regulatory hurdles to working with communities. Community members will have different values and goals from a data science team. Many such communities are used to hearing empty promises from outsiders without any future payoff, and in fairness, many researchers see communities as sources of data, not as potential equal collaborators. Appropriate data about and from these communities might be hard to collect. It will take time to develop trust with communities—and yet this often is at odds with the pace of data science projects where pipelines, models, analysis, results, and dissemination of results are attempted as quickly as possible.

But all is not lost! There are plenty of positive examples to draw from (Baumer et al. 2017; Dencik et al. 2019; Hirsch 2011; Saxena et al. 2020; Taylor et al. 2019; Zegura, DiSalvo, and Meng 2018). In general, to work with communities as we defined above, you should educate yourself deeply about their particular context as well as explicitly communicate about the communities' goals and your intentions.

Trust is built on mutual understanding and mutual respect. Too often, participation with communities does not seek to deeply involve people in the work. Urban planner Sherry Arnstein (1969) described an eight-step ladder of participation, from nonparticipation to complete community control over the project. These degrees of participation can be thought of as a data science team informing, consulting, deciding together, acting together, and supporting the communities' own project (Wilcox 2004). Consultation with community members is one stage and may be appropriate for many projects. But mere consultation without empowerment about decisions is not very participatory. There may be ways to *support* communities as they lead in formulating the questions and designing the projects. Collaboration with communities on their unique needs is one way to help solve *their* specific problems. Data science teams can train community members and invite them to be part of analysis, empowering them with decision-making capabilities and skills. For example, these mixed teams could provide the opportunity for community members to decide which features may be used for predictive power in a machine learning model. Finally, it is a great idea to think about an iterative process and provide opportunities for community members to be invested in the project.

Action research provides a different set of models for relationships between communities and academic or industry professionals (Hayes 2011; Reason and Bradbury 2005). Action research generally involves *mutual* education, in which community members educate researchers about the history, needs, and goals of the community, while the researchers educate the community about technical or policy opportunities. In these projects, each party commits itself to the success of the other party. While action research is a very well-respected way to "do science," it uses different "criteria of merit" from so-called objective scientific paradigms. It is less about formal hypothesis-testing and more about inquiry, discovery, and mutual aid to create new scientific understandings. Data science workers who engage with a community in this way are likely to develop extensive domain knowledge, which they can then apply to making a better-informed, more effective, and more sustainable data science solution that meets the needs of the community.

With these approaches in mind, consider the following questions:

- Think about a community you want to work with. Who are they? What problems might they have? How would you approach them to talk about these problems? How might a data science approach help?

- What questions would you want to ask your community partners? How would you include community members in your process? How would they like to be included? What would help empower community members? What would success look like for you? For your partners?

Working with People's Data

Shoshana Zuboff (2019) raised questions about how commercial and government entities collect and use data about people under the general topic of "surveillance capitalism." Stuart Thompson and Charlie Warzel (2019) have published a series of articles about the surreptitious collection of personal location data through people's smartphones and the sale of this data to third parties. Earlier scandals involved unauthorized collection of personal data for unacknowledged use by political campaigns. Certain national governments now collect increasing amounts of personal data about their citizens and residents and then use analyses of this data to target people for potentially unjust treatment or outright repression.

Data science is deeply implicated in these practices and in these potential intrusions. Data science applications may collect data that people do not know they are giving up and analyze this data to infer things about people that they have not given permission to be inferred or distributed. Or people may not be aware that their data is openly available to data scientists. For example, in a recent survey of Twitter users, the majority felt that "researchers should not be able to use tweets without consent" (Fiesler and Proferes 2018).

- How should other people be able to use your personal data? What uses would you consider to be unacceptable?

- How do you think about your responsibilities with respect to other people's data? Which uses do you think are okay? Which uses would you find unacceptable?

- Which applications on your smartphone have access to your contacts? Which applications need that access? Do you know how to deny access to the applications that you think should not be able to use your contacts data?

- Which applications on your smartphone have access to your location data? Which applications need that access? Do you know how to deny access to the applications that you think should not be able to use your location data?

- When you are creating a data science application, how do you decide which personal data you actually need? What protections do you put in place to defend the privacy of the data that you are collecting? How could those protections be breached?

Conclusion

Human-centeredness refers not only to the people whose data is being analyzed but also the data scientist, the data science team, and other stakeholders around the data science problem of focus. There are different forms of teamwork and collaboration.

Asking questions about whom we work with and for allows us to think about people and their power in relation to a data science system, to each other, and to a data science team. Employing Value Sensitive Design can add analytic precision to our analyses, especially regarding direct and indirect stakeholders. Some roles in data science work appear to conflict with other roles, raising difficult questions about who has responsibility and authority to define and design some of the potentially problematic aspects of data science systems.

Data scientists rarely work alone. Team members may come from backgrounds similar to data science or from other backgrounds and disciplines. Considering issues of disciplinary difference while working on such a team allows us to collaborate more smoothly with people from different backgrounds. Importantly, there are distinct roles within data science teams sometimes populated by folks with different professional and academic backgrounds, although many data science teams include "pi-shaped" people trained in two or more fields. Data scientists need to learn how to negotiate these issues to address problems.

Collaboration with AI systems presents unique challenges of power, control, action, and feedback.

Working within organizations also presents unique concerns. We explored the notion of stakeholders from VSD and their particular values through which they see any given data science problem. Objectives, agendas, and outcomes of different stakeholders may intersect or diverge from that of the data science team, and thus it is important to listen to and include the constellation of values surrounding a project.

Data science often seeks to do social good in working with vulnerable communities. This presents challenges for the data scientist as such communities might have legal, regulatory, or trust issues that prevent effective collaboration. Data scientists might be tempted to address problems with what is easily available or easily quantifiable—which might not be the most suitable approach for the problem. We have suggested some directions that provide contextual, participatory, and community-centered approaches to working with communities.

Recommended Reading

"Community-Based Participatory Research: A Guide to Ethical Principles and Practice." 2012. Centre for Social Justice and Community Action, Durham University, and National Co-ordinating Centre for Public Engagement. https://www.publicengagement.ac.uk/sites/default/files/publication/cbpr_ethics_guide_web_november_2012 .pdf. A short yet thorough guide to ethical choices in working with community groups.

Degani, Asaf. 2004. *Taming Hal: Designing Interfaces Beyond 2001.* New York: Palgrave Macmillan. This book presents an approach, mostly based on software engineering and formal methods, for describing, analyzing, and identifying potential design errors in human-automation interfaces.

Eubanks, Virginia. 2018. *Automating Inequality: How High-Tech Tools Profile, Police, and Punish the Poor.* New York: St. Martin's. A gripping set of stories of how technology affects human rights.

Olson, Gary M., Ann Zimmerman, and Nathan Bos, eds. 2008. *Scientific Collaboration on the Internet.* Acting with Technology. Cambridge, MA: MIT Press. This excellent and comprehensive reference includes more than two decades of study of how scientific groups collaborate over the internet.

Wilcox, David. 2004. "A Short Guide to Partnerships." Partnerships Online. http://www.partnerships.org.uk /part/index.htm. Short, readable guide to working in communities, written from the perspective of community leaders.

"Working with Local Communities." 2017. National Co-ordinating Centre for Public Engagement and University of Bristol. https://www.publicengagement.ac.uk/sites/default/files/publication/working_with_local_commu nities.pdf. This short report, targeted at academic researchers and institutions, has a good checklist for thinking through the issues of collaborating with community partners.

8

Storytelling with Data

When one of us (Gina) was an undergraduate economics student, the late Caroline Dinwiddy taught her econometrics course. "Don't make the data speak," Dinwiddy warned the class. "Let the data speak to you."

At first glance, telling stories may seem to have little to do with data science. One is the practice of rational and careful work, years of study, statistics, and evidence. The other? It is about emotions, things we learn at home and from our communities and families. Storytelling is something we come to primary school loving and knowing how to do.

And yet data does not speak for itself. Data does not simply tell its own story. It is our job—your job—to tell the story of our pipelines, our data, our models, and our results—and what they mean for others—responsibly, accountably, and carefully. To be successful in data science you will need to learn to communicate something you discover to others. Before you can tell that story, it is important to think about the audience, how they will hear it, and what you want to accomplish. Do you want your audience to explore your results? Play with them? Is your audience relying on your analysis to make decisions? Does your audience need persuading or convincing? This kind of reflective self-knowledge can help to shape your story.

Our goal is to help you communicate more effectively as an advocate for your data science project. Studies of data science practitioners show that they find communicating their results and persuading people to use them to be one of the biggest barriers they face. Scott Berinato (2019), a senior editor of the *Harvard Business Review*, named three common pitfalls in communicating data science. The first is the "statistician's curse," when a data science practitioner thinks that brilliant models and analyses are sufficient, therefore attention to the design of visualizations and presentations is not important for communicating with others. The second pitfall happens when busy business leaders, armed only with visualizations from their data science teams, cannot effectively communicate with others. The third is "the convenient truth" (also sometimes known as "confirmation bias"), when people look at data science results only to support their earlier beliefs, rather than exploring what the results might mean.

Each of these pitfalls shows how the responsibilities of data science practitioners do not end with their analysis. We argue that data science practitioners should think about the entire process of data collection and production as one that has communication at its core. Communication practices are key data science practices. A human-centered approach to data science takes seriously the call to help others to think deeply about the results and limitations of data science. It means that practitioners think about communication as well as technical or quantitative content, and that they anticipate how communication shapes their work. They can then be ready to translate their work across multiple audiences and contexts (Neff et al. 2017).

We believe that data science presents an opportunity to connect with other people. Paying attention to how we communicate about data science pays off in the impact of our work. That attention benefits us, the clients and communities we work with, and the people impacted by our projects (see chapter 7). Stories are a useful way to frame that communication process. If we understand *communicating data science* as *telling stories*, this knowledge will encourage us to think about what resonates with our audiences. It helps us to consider urging people to act on results—whether that is thinking more deeply about our results or persuading them to make a decision or take action because of them.

Using storytelling with data as the way to talk about data science communication helps us to remember that communication is a process that involves others, and that the results of our projects are not the end of our job but in some ways only the beginning. Storytelling can help us remember that others want to know "what happens next" when they listen to a story, and about the importance of "understanding the structure and mechanics of narrative and applying them" to data visualizations and presentations (Berinato 2019).

Storytelling reminds us that there are many ways to end a story, that cultures have very specific ways of telling stories, and that we are accountable for our interpretations of stories. Bringing this approach into data science requires us to take a more holistic view of how our results "live" in the stories that other people tell about them, and our responsibilities to those results. Poet Adrienne Rich (1986, 34) wrote about the impact of our work that lives beyond our intentions:

We move but our words stand
become responsible
for more than we intended

Storytelling reminds people in data science that our projects, our pipelines, our models, and our results can have a "narrative arc," a format that helps others think about what is next. Listen to the fifteen-minute TED Talk of any great data scientist (such as Hans Rosling) and you'll hear powerful stories that data can tell, and the stories that people tell with and around their data projects.

- Think about the last time you watched a TED Talk (if you haven't watched one, go to www.ted.com and find something that interests you). What made it work? Did the speaker use stories or narratives? If yes, did the speaker do that to grab your attention, teach you how something happened, instruct you how to do something, or motivate you to take some form of action? How could you adapt what they did to strengthen your own communication skills?

Why Stories Matter to Data Science

We started this book by evoking the hope that it would be a guide for you to change the world as a data scientist. Many times, data science results may appear to be one end of the data science cycle. However, the results stage is only the beginning of a conversation that people in data science have with others. The goal for us in teaching a method for human-centered data science is to remember that there are people involved in making data science. There are people who use our results, people whose data powers the science, and people who deal with the consequences, intended or not, of decisions and actions that are based on the analysis of data.

Without effective ways to teach and reach these people about our projects, our work has limited impact. Storytelling is more than simply creating powerful and memorable images to help people understand and remember our work. Storytelling reminds us to use language that people are comfortable with to explain our work, justify our choices, and teach others what we know. We can use the metaphor of translation here. People working in data science speak a specialized language, one that you have learned through this book and in your other data science education. Translation scholars talk about translation as involving both transformation and transportation (Maitland 2017). Translating what we know to others who do not speak that language does not mean dumbing it down. Rather, it means figuring out how to solve the dual problems of transformation and transportation. We need to *transform* our description of our project until it makes sense to our audience. Simultaneously, we need to *transport* our description from our own world of data and algorithms to someone else's home terrain—which may be project management, grant making, administration, sales, medicine, or science, as examples. The goals in our world may be different from the goals in our audience's world. If we want to be effective, we need to speak to their expectations and their needs—their world—even as we explain what we did in our world. In any communication project, that means understanding the context of when certain things are acceptable or appropriate (e.g., the way we speak in a classroom is different from how we might talk in a café with our friends). It also means understanding our audience (e.g., we speak one way to our grandparents and another way to children).

One of the authors (Michael) had successfully presented the story of a data science analysis to his work group. When Michael gave his executive sponsor a copy of his presentation, the sponsor rejected it, saying, "I need materials for senior executives. They just want a table of numbers. Make sure it has exactly what is most important. We only get one slide, and only one chance." Despite the work group's interest in the history and drama of the project, Michael had to *transform* the story into a single table to *transport* the story from his research world into the sponsor's executive world.

Chris Wiggins, the head of data science at the *New York Times*, sees communication skills as one of the crucial things data scientists need to be able to do:

Communicating the results to everybody else in a clear and concise fashion, that is definitely something I require of people that I hire in the data science group. I want data scientists that I can drop in a room with somebody from product or marketing or somebody who does not speak calculus. . . . We want people who can communicate in a clear and concise fashion to the rest of the org. (Wiggins 2019)

The point of storytelling in data science is not to make up fictions. Far from it. What we are encouraging you to do is to try strategies that help you communicate with others in a way that maximizes the impact of that exchange—both for your goals and for your audience's needs. Learning to identify what is meaningful, interesting, and surprising in a data science project is a process of storytelling. For example, "I expected to find this; instead, I found something else." Consider the point in a story when the narrator says, "And all of a sudden." In just saying the phrase, the author has created an expectation that something follows and that there will be some form of change. Wait for it. Get ready for a new piece of information that might change your mind!

Our well-practiced expectations for how stories work create a model for examining our data science to find what is important for our project and what will seem important to others. This comes with responsibility. Storytelling is a framework for thinking about data science communication and about how stories shape how others hear what we are saying. It is also a way to push yourself to think through what is important, unusual, interesting, unexpected, or novel about your project. Ask yourself what stories you can honestly and responsibly tell with your results.

- Think of a recent project. Try to make really short stories that tell different things about your work. Can you turn that project into an inspirational story (go and do this)? Or a cautionary story (*don't* do this)? Can you turn that project into a description of how to do the project (here's how you can do this)? Can you add a character who is not you and make it a story about that character who did something to make or use your system? These kinds of short narratives are good practice for writing a story for your project when the story will really make a difference.

The Importance of Stories in General

Human brains are hardwired to learn from storytelling, as the neuroscientist Antonio Damasio (2012) has found. It's an age-old technique to pass on wisdom from one person to another. A dry recitation of facts is likely to be forgotten quickly, but we perk up and pay attention when we hear a story with compelling characters and events.

This is especially crucial when dealing with large amounts of data, which the human brain does every day. Based on some estimates, we are bombarded with millions of bits of information every second, and our conscious mind only records maybe forty of them (Wilson 2004). The human brain has evolved to select and retain only the information that's most important for our survival. How? Our brain creates meaning from the complex information around us and casts it into a narrative. Storytelling evolved as a method to enable us to learn from experience without having to live through it. It was necessary for our survival. This may explain the compelling nature of a story—why it grips us all so dramatically, why it catches and holds our attention. A story is a simulation that can save our lives.

Thus, when we as data scientists attempt to communicate the information behind large amounts of data, it is critical to tell a story so that our audience is primed to pay attention, understand, and learn from our message. As we mentioned earlier, it's not enough to simply "show the data." We need to shape the data into a compelling story—without embellishment and with honesty and responsibility. We also need to make the story "fit" within the audience's needs and expectations. A colleague of one of the authors (Michael) sometimes visited the Pentagon building in Washington, DC, to deliver reports. He wanted to make the most of his opportunities to persuade senior officers to support his projects. His main way to do this was to deliver a literal elevator pitch to generals at Pentagon headquarters. To be fully prepared, he measured the duration of a Pentagon elevator ride: twenty-three seconds. Note that stories don't have to be long to communicate their point. This story has fewer than seventy words and could perhaps be conveyed during a ride in the Pentagon elevator.

- Go back to the short stories that you created earlier. How could you make the short stories *even shorter* so that you could deliver them during a twenty-three-second elevator ride? What would you cut out? What does that tell you about the absolutely necessary, irreducible concepts in your story? The strategy of making it much shorter can sometimes help you to see what is essential, what is not, and what is essential *for this audience*. The answer may be different for different audiences.

Visualizations as Storytelling

Visualization is an intrinsic part of data science communication. As you seek to communicate the story behind your data, you'll often find that visualization—or what we describe in more detail as the process of visual storytelling—is one of the most effective methods to do so.

As described in chapters 2 and 4, visualization is a powerful and effective way to convey large amounts of information to the human brain. Nevertheless, it's not easy to create effective visualizations. It is a skill that must be learned, and one that combines both analytical abilities (i.e., the ability to analyze and classify data, identify data types, and clean and sort the data) and design skills, which is a field all on its own.

The most effective visualizations tell stories.

But visualizations can be misused. Data can be inflated or deflated, taking advantage of biases in the human visual system to distort the underlying truth in the data being visualized. This may be deliberate—where the creator seeks to amplify or distort a message—or inadvertent—where a designer may simply be ignorant of the best method for a particular audience. In either case, bad visualizations tell a story about results that fits neither the data nor the audience's goals and needs.

In the field of information visualization, there has been a great deal of interest in visual storytelling. Cole Knaflic's *Storytelling with Data* (2015) contains many excellent examples. Other researchers (Segel and Heer 2010) describe patterns in narrative visualization and classify visual stories as "author-driven" or "reader-driven." Author-driven approaches present a linear path, include explanatory text, and have no interactivity. Examples might include videos or slideshows. Reader-driven visualizations do not prescribe an order in which to view them, include less explanatory text, and have a great deal of interactivity. Tools such as Tableau or web-based interactive visualizations might fall into this category. If your ultimate goal is to communicate, an author-driven approach will work best. If your goal is to allow people to explore your data, then a reader-driven visualization may be more effective.

Hybrid approaches can also be taken. Edward Segel and Jeffrey Heer (2010) discuss a particularly interesting structure that is often used in data journalism: the "martini glass." It begins with an author-driven introductory visualization that may contain explanatory text and a more static view of the data. Once the reader has finished this part of the

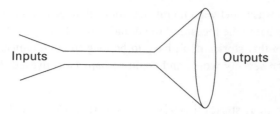

Figure 8.1
Martini glass metaphor for data journalism.

narrative, the visualization opens up and becomes interactive and more reader-driven, allowing free exploration of the data. Segel and Heer find this hybrid approach takes the structure of a martini glass, with the initial narrative forming the more linear stem and the interactive portion the more expansive bowl.

How Stories Work

A story is an ordered sequence of events or scenes, with a clear path leading through these scenes. Order may correspond strictly with linear time, or flashbacks and other techniques may be used to present ideas in a nontemporal order, using a narrative logic that reveals ideas as they are necessary to understand the storyteller's presentation. Nevertheless, certain techniques tend to be common to all stories.

- Stories have a beginning, middle, and end.
- Stories are filters. They focus only on a particular problem or issue that needs to be followed. Thus, they accomplish exactly what the human brain does: ignore all distracting elements.
- Stories grab our attention. They may do this through surprise, curiosity, anticipation, a reason to care, or the presentation of a problem that needs to be solved.
- They often contain explicit, well-defined characters, although the characters may be implicit, unknown, or may simply be the reader.
- Something has changed by the end of the story. Traditionally, the main character experiences an emotional change or learns a life lesson. However, in data science stories, what has changed may be the viewer's understanding: the message received, the knowledge imparted, and perhaps future decisions made.

Other effective storytelling techniques include narrative flow, surprise, hooking the reader, and the power of repetition. Ask yourself the following questions about stories to improve your data science communication:

- Think about a story that you recently read, heard, or watched. What made it "work" for you? How did it keep your attention? Were there moments that surprised you? How did it prepare you for dramatic changes? Did the story "telegraph" its ending so that you could enjoyably anticipate it? Did it turn out the way that you expected? How could you use these techniques in your own storytelling for data science?

Case Study 8.1
Examining the Ethics of Visualization Design
Jaime Snyder, University of Washington Information School

"Most information visualizations are acts of interpretation masquerading as presentation . . . they are images that act as if they are just showing us *what is*, but in actuality, they are *arguments made in graphical form*." (Drucker 2014)

"All visualization research, no matter how superficially apolitical or trivial, has a moral character." (Correll 2019)

Leading up to the US general election in November 2016, widespread models projected a likely (but not guaranteed) victory for Hillary Clinton. Post-election, after Donald Trump's victory, the media focused on how and why these models could be so "wrong." Statistical phrases such as "confidence interval" and "uncertainty" popped into public discourse to explain that many of these models had, in fact, left open a clear statistical possibility of a Trump presidency.

A few weeks after the election, I attended a conference for visualization professionals. At the gathering, conversation turned to the visualizations of those predictive models. I was struck by what appeared to be consensus among attendees that neither the visualizations (nor the designers of those visualizations) were to blame for creating false expectations. Rather, the fault lay with the general public's lack of statistical literacy. I was not sure that I agreed.

In the context of data science, the visualization design process is typically treated as a technical endeavor driven by scientific standards of objectivity (Burri and Dumit 2008; Lynch 1988; Pauwels 2016). This is reflected in the perspectives of the data visualization professionals I encountered at the conference. In contrast, the field of *visualization studies* focuses on examining the ethical, social, and communicative tensions embedded in the creation and use of visual representations of information, data, and knowledge. Using participatory design, ethnography, and other forms of qualitative research, this branch of the study of data science (Neff et al. 2017) explores questions such as: Is it the ethical responsibility of designers to foresee misinterpretation stemming from lack of expertise? Or, in our current data-driven world, is it the moral imperative of the general public to acquire the visual and statistical literacy to properly understand probabilistic representations and to master possibly counterintuitive interpretations?

Katie Shilton, a social scientist who studies values and ethics in information systems, explains that "values held by designers affect how information technologies are imagined; how systems handle data, create categories, and draw inferences; and what affordances are available for user interaction. All of these decisions affect the social consequences of emerging technologies." (Shilton 2013, 375).

Values, in this context, refer to "what a person or group of people consider important in life" (Friedman, Kahn, and Borning 2008, 69). The values of an individual or a community can greatly influence which design problems are given priority (Le Dantec and DiSalvo 2013; Fleischmann 2013; Nissenbaum 2001), how tools are provisioned (Koepfler and Fleischmann 2012; Snyder and Shilton 2019; Snyder, Shilton, and Anderson 2016; Woelfer et al. 2011), how resources to solve those problems will be allocated (Fleischmann and Wallace 2005; Friedman, Kahn, and Borning 2008; Snyder 2017), and importantly, the metrics by which the credibility of information artifacts, like the 2016 general election visualizations, will be evaluated (Snyder et al. 2019).

As I write this, the world is gripped by the COVID-19 global pandemic. Messages from local, state, and federal government call on us to "flatten the curve," making reference to charts created by public health researchers showing steep upward arcs projecting catastrophic impacts on our healthcare system if citizens do not act swiftly to curb the spread of the coronavirus (Institute for Health Metrics and Evaluation 2020). In many ways, this can

(continued)

Case Study 8.1 (continued)

be seen as an exemplar of *ethical visualization*: a graphic representation of critically important information made available to a wide audience with efficiency and clarity. However, as Correll (2019) notes, it is also crucial to examine the underlying data of representations like this. In this case, the simplicity of the "flatten the curve" message belies troubling social and economic inequities related to (1) which citizens *can* stay home without jeopardizing their income and (2) which already vulnerable groups are disproportionally burdened by these drastic measures.

While questionable and unethical displays of data can certainly be the result of intentional deception, they can also result from a misalignment between the designer's intentions, the context in which the visualization is encountered, and the specific assumptions and prior knowledge of audience(s) who interact with the graphic (Hemsley and Snyder 2018). These factors encompass the widening range of people with access to the tools and data needed to create professional-looking data visualizations. They also provide a means to understand the multiple phases of sensemaking that occur when visualizations are posted on and across social media platforms.

Ultimately, we need to understand the ethics of visualization as a complex and contextual process of sensemaking. This understanding not only provides methods for assessing the credibility of visual representation of information, but also reveals opportunities for data designers to empower individuals to engage with data on their own terms, to improve data literacies, and to recognize that the evaluation of data credibility in real life is multidimensional.

Storytelling as Communication

Storytelling is one powerful way we learn to connect with others. People use many different ways to communicate. But we use storytelling here for two reasons. First, in business, data science storytelling has become a common way of communicating results. Second, we use it to highlight that communication is a key data science skill—and one that is often overlooked in data science training. One of the authors (Gina) followed energy engineers in their work with architects in modeling the potential energy consumption of the buildings that they were designing to present clients with options for more energy-efficient buildings. She watched industry leaders—those energy engineers who were high in demand—early in the process. Their analysis was sophisticated, their models complex. And yet much of what they did in practice was work with a team of people who were not trained in data analysis to help them figure out what was important about the design and to help them understand the impact of their choices. Time and again we see this in watching how expert data science practitioners work—with careful attention to the practices of communicating, helping others understand what went into their models, and clearly showing how their choices may have influenced the final results.

This does not surprise people who have observed data science in practice. Information science scholars Samir Passi and Steve Jackson studied the work of data science teams in business. They show how business leaders trust the work of their data science teams in no small part because of the work that the teams do with negotiation and translation of the results. These leaders trust data science largely *because* of this work of communication—which is just the opposite of the "statistician's curse." They show how data science teams brought their audience back into the story, literally, through narration and storytelling, often increasing trust in the results (Passi and Jackson 2018).

• Go back to the TED Talk that you analyzed earlier. Watch it again with a critical eye, using the ideas that you have just read about. Can you analyze how its story or stories worked? How did the storyteller build suspense? What were the dramatic parts? Were there dynamic surprises? How did the story keep you engaged with the topic or the character (or both)? Would you have told the story differently? If yes, what would you do differently, and why? If no, what made it perfect the way it is?

How to Talk about Data Science

Remember from the discussion of disciplinary bias in chapter 7 that it is extremely important to pay attention to your own use of language and to speak to your collaborators using the terminology they are comfortable with. Often the secret to a successful collaboration is as simple as word choice. In one project, Michael interviewed people who managed the process of designing ground-truth labels to apply to data records. Many informants described challenges in developing a vocabulary of labels that simultaneously made sense to the labelers *and* to the client. Ultimately, the client's need to have understandable labels became the most important factor—even though the labels were applied internally to the data science team and were only *temporarily* visible to the client. Once the labeling vocabulary had been agreed on, the client was ready to trust the work of the team (Muller et al. 2021).

How to Talk about Human-Centered Data Science

Again, this is a story of language. One of the authors (Cecilia) says that when she first joins a new group, particularly in the physical sciences, she doesn't begin by saying, "I will use an ethnographic and contextualized approach to understand your problem," even though that *is* what she does. The essence of human-centered data science is centering the human. This means putting your audience first. So, as a new data scientist in a group of experts in their particular domain, you first have to *listen* to understand their needs, rather than demonstrating your own expertise. Starting with a "requirements document," as many software projects do, may not be successful (as a matter of fact, there are many cases where starting a project by asking managers and experts for their requirements has led to failure). The issue is that very often people who are not data scientists may not know what they need. Of course, neither do you as the incoming data scientist. On the other hand, some requirements may not be technically feasible. Listening is required of both sides. Once you understand the audience's needs and come to a mutual understanding, then you will be in a much better position to choose the right strategies for data cleaning, feature engineering, modeling, and evaluating your outcomes.

Visualizing Is Part of the Story

As we mentioned in chapter 2, the human visual system is the highest-bandwidth channel into the human brain (Ware 2020). Data visualization, as we described in detail in chapter 4, is an incredibly powerful way to get your story across. One way to tell the story of data is to develop *vignettes*. This term from literature describes short, impressionistic scenes or sketches that highlight some key characteristic. Such short text descriptions can be illustrated by a particularly evocative screenshot or graph that helps make your point. Vignettes can be used to tell the story of how users navigate a service or tell stories of other human-centered data science techniques. They are verbal ways to help others picture and understand the data.

Vignettes can also work in video form. Michael used a video recording of a user interface failure to motivate a development team to make changes. To begin, the recording generated sympathy and admiration for the user, who had an advanced degree in mathematics. The recording showed her receiving a series of error messages that made no sense. After yet another "beep" error sound from her workstation, she replied to the user interface, "Go beep yourself," and that was the end of the vignette. Because the short video had established her technical expertise and authority, the developers were urgently convinced that they wanted to make the design better for such an intelligent (and funny) user.

Without understanding your audience, it is tough to imagine how they will hear your story. Consider the example from Scott Berinato (2019) of a business leader who wants a data science team to generate a visualization for an upcoming presentation. Here, the audience (the business leader) is someone who needs to convince others to make a needed change. Now imagine the businessperson's presentation is to a group of clients who are considering buying a new product or service. Or what if the presentation was being made to others in the company who were looking to make a decision? The goals of the presentations would be different. Creating a story with the data means thinking about what needs to be conveyed to the audience, who will be telling the story, and how they need to tell it effectively.

In sum, communication is a process, much as a story is a collaboration between storyteller and listener. Learning to think of data science communication as a dialogue between people with data science results and others helps us to see the importance of creating that link.

Risk Management for Business Leaders

Being human-centered in data science means focusing on one's audience, and that matters when it comes to business leaders. It can be challenging to walk nonexperts through the process of data analysis. It can be tough to present results in a way that doesn't "oversell" them or shares the level of uncertainty or contingency in models. It can be hard delivering bad news or talking to bosses.

One powerful way to tell the story of the importance of human-centered data science is to describe it in terms of risk management. Ethical considerations may constitute an up-front expense, but history has shown that a shortsighted focus on immediate costs can impact the ultimate return on investment. The risks of ignoring human or ethical concerns at the beginning of a project can lead to huge downstream costs. We have presented many such examples in this book, particularly in chapter 5. We have all successfully argued to managers and business leaders that a human-centered approach to data science is simply good risk management. Learning to manage the expectations of different

Case Study 8.2
Building Visualization Tools for Experts and Scientists
Alexander Lex, University of Utah

Interactive data visualizations have become part of our daily lives, thanks to the news media publishing informative, engaging, and interesting data visualizations on a daily basis. The COVID-19 pandemic specifically has been accompanied by a plethora of dashboards communicating every facet of the development of the crisis. Data visualization, however, is equally important in a different scenario: when experts need to analyze scientific data.

I develop visualization techniques for these cases. Since each technique is tailored to a specific analysis problem by domain scientists, my group follows a user-centered design process. Traditionally, we emphasized immersion in our collaborators' environments: students working on developing a visualization technique would embed themselves in the research group of a collaborator for multiple months, participate in group meetings, and observe their workflow. Since this is a highly time-consuming endeavor, we recently developed a method to fast-track this process: we now frequently use "creative visualization opportunity" workshops (Kerzner et al. 2018), that can last from a half day to two days, to kick-start a collaboration. We found that bringing all stakeholders together and having them go through thought experiments, unconstrained by realities, helps to understand the problem quickly and to fast-track our research process.

After that initial phase, we start developing the visualization tool. We first use sketching to explore alternative concepts, but we quickly move to prototyping. In every case I've seen, data is messy and full of aspects you didn't consider, so even the most sophisticated design can fail if you try to implement it with real data. To avoid this, we develop prototypes in code with real data as soon as possible. A well-known pitfall occurs when no real data is available yet. I've encountered this myself: I once spent ten months working on a visualization tool for a hypothetical dataset that never materialized. Avoid projects where you can't get the data.

The job of a good visualization designer is to give users what they need, which might not always be what they want. Collaboration partners often have ideas on how to visualize their data themselves and are primed by domain conventions. It is usually helpful to question these conventions and to try to understand the problem they want to solve instead of the visualization they want you to design. For example, red-green heat maps are a common tool for analyzing biological data, and I've been in many situations where a partner has asked for an "interactive heat map" and insisted on a red-green color scale because "that's how we do it." However, this rarely is the solution, as we know from vision science and visualization theory that a lot of people are red-green color blind and hence can't read these heat maps. A visualization researcher should not become a software engineer to fulfill the partner's wishes, but instead takes the role of a "data counselor," helping their collaborators focus on what it is they want to understand and which data, in which form, they need to get there.

•

One example of such a collaboration is when we worked with a team of psychiatrists studying genetic factors of suicide. They used large genealogies to identify suicide cases that might share a genetic variant but were struggling to see and understand their data because of the size of these genealogies and the many attributes about the individuals they needed to consider. Using our process, we developed Lineage (Nobre et al. 2019), a tool they can use to explore families with hundreds of members and attributes about the individuals who committed suicide. They now leverage the tool to screen for individuals where genetic causes likely played a role. For example, they can look for relatedness, associated rare psychiatric conditions, and age to narrow down the number of cases to study.

(continued)

Case Study 8.2 (continued)

As computer science researchers, we care about helping with specific, high-impact domain problems, such as studying suicide, but of course we also hope to develop general knowledge that can be applied in different contexts. It turns out that these specific techniques can frequently be generalized. In the case of Lineage, we used the ideas to develop a general-purpose network visualization technique (Nobre, Streit, and Lex 2018) that can now be used to explore datasets as diverse as social networks or gene regulatory networks. And obviously, designing visualization techniques that can be used to solve high-impact science problems is a rewarding endeavor on its own.

audiences while advocating for the work of data science is one of the key aspects of working in data science in industry.

What Drives Policymakers

The needs of data science for policymaking and government present us with unique challenges. Governments need to be accountable to public scrutiny. Working to make public-sector services better through data science and innovate within government is a great task to undertake. Governments need to be able to explain how and why certain decisions were reached. Public-sector organizations serve the public, are accountable to them, and should be nondiscriminatory. They must be able to explain to policymakers and the public both the process and the outcomes of decisions, including those that involve machine learning algorithms. Ethical action and public safety are important for public-sector organizations that need to ensure that people who are socially or economically vulnerable do not bear a greater burden of their choices. Public-sector organizations have a dual responsibility to be accountable to citizens and to exercise care for the public trust.

All of these have implications for doing data science. The UK's Alan Turing Institute collaborated with the UK Office for Artificial Intelligence and British Government Digital Service to produce a comprehensive guide to best practices for data science and AI implementation in the government sector (Leslie 2019). They recommended a series of steps to ensure that public-sector data science projects meet the unique needs of the public sector, including first assessing the impacts on the most vulnerable populations, having a clear plan for communicating with affected communities, understanding how the project will be implemented, and clearly mapping the roles and responsibilities of people in the process. In other words, human-centered data science *is* the best way to build projects in this sector. "To provide clear and effective explanations about the content and rationale of algorithmic outputs, you will have to begin by building from the human ground up. You will have to pay close attention to the circumstances, needs, competences, and capacities of the people whom your AI project aims to assist and serve" (Leslie 2019, 57).

For data science projects to work for the public sector, people working in data science must (1) maintain strong regimes of professional and institutional transparency, (2) have a clear and accessible governance process for projects, and (3) ensure projects are designed so that they can be audited and explained later (Leslie 2019, 36). "Translating" the model "back into the everyday language of the humanly relevant categories and relationships that informed the formulation of its purpose, objective, and intended elements of design in the first place" is essential for the success and safety and public acceptance of these projects (Leslie 2019, 60).

Working with Experts

As we described in chapter 7, in data science we borrow terms from software engineering to describe an expert as a domain expert or subject-matter expert. People in data science rely on others who understand the particular situation, context, data, or problem at hand to help them design better human-centered data science projects. To communicate with these experts, we often rely on storytelling to get our point across in a nontechnical way, and we may ask them to engage in storytelling to educate us (Muller 2001). Analogies can also be extremely helpful.

Working with scientists can be both extraordinarily rewarding and extremely challenging. Scientists are accustomed to questioning methods and statements, and they are often accustomed to solving problems in a particular way. One of the authors (Cecilia) worked with scientists as they were transitioning to dealing with larger and larger onslaughts of data in their normal workflows. She worked with groups from astrophysicists to biologists to particle physicists who deal with dozens of gigabytes of data per second. Although the particle physicists were working with the largest data magnitudes, she found commonalities across all groups. All project teams faced the need to write completely new software, invent completely new algorithms, and completely change all their work processes as the amount of data they had to process grew. The way each team member worked on a daily basis changed, sometimes radically. The magnitude of the raw data size didn't matter as much as the magnitude of the increase. This led her to define the term "big data" not in absolute terms but in a human-centered way: big data is 100 times greater than the amount you are used to dealing with. She tells a human-centered story to give people a visceral understanding of the situation (since humans don't have a very good intuitive concept of the difference between a gigabyte and a terabyte): Imagine that your job requires you to read two books per week. Time-consuming but manageable. Now imagine that you need to read 200 books per week. Suddenly, everything has changed. You need to figure out a completely different way to collect the data from those 200 books. That is the human impact of big data.

In engineering and business, the argument can often be made by focusing on the bottom line or underlying goal of the project group. Here, the language should focus on business or engineering needs. Engineering and business experts will often listen closely if you describe how your approach will meet their needs and how a human-centered approach is prudent risk management. If you are listening to their needs, you are on your way to project success. A short, focused message can be very effective—remember Michael's colleague who timed the Pentagon elevator to deliver an "elevator pitch" in twenty-three seconds.

Working with Communities

Working with the people our data concerns, and with the people potentially affected by our projects, is an important part of human-centered data science. Consider the following

Case Study 8.3
Knowing Why in Addition to What: Designing Explanations in Data Science
Q. Vera Liao, IBM Research

As data scientists build models that are interacted with by people in the real world, they may notice that users frequently demand *explanations*—for example, asking *why* the model made a particular prediction. Sometimes the explanation is a more important part of the user experience than the prediction itself, as we realized in our work.

A group of us were working on a project to create an AI tool that predicts patients' risk of adverse events, such as unplanned hospital admissions, so that physicians could provide appropriate care and resources for high-risk patients. Very quickly we realized that providing risk-prediction numbers about a group of patients was essentially useless without explanations, because physicians were reluctant to pay attention to these numbers. As we talked to some physicians, they said they "just want to know why." There were at least two reasons for them to ask the *why* question and expect to see explanations.

First, physicians wanted to understand the model's predictions in order to trust it and, in necessary situations, to cautiously bypass the model and rely on their own judgment. This was especially true when physicians were just starting to use the tool. Seeing even one prediction that did not align with their expectation about a known patient could deter them from trusting and continuing using the tool. However, if they were given an explanation—for example, the model made a questionable prediction for a given patient based on personal data that might not be up-to-date in the database—the physician would have not only understood and forgiven the model's mistake, but also appropriately chosen to pay more attention to the patient.

Second, physicians wished to learn from the explanation, to understand what features of a given patient might be the critical factors that put them in a high-risk group. Such information was highly valuable for physicians to choose follow-up actions for risk intervention. For example, if the model explained that it predicted a patient to have high risk because of a past event in the patient's medical history, which the physician might not have been aware of, this information would have been very useful for designing a care plan to avoid similar events. In our case, explanations are in fact a more central element of the user experience than the predictions for a clinical decision-support tool. Our final design of the tool therefore featured a visualization that explained which were the important features for a given patient that contributed to the patient's predicted risk, especially highlighting features that were useful for intervention planning, such as the patient's medical history and socioeconomic factors.

There was also another side of the story about AI explainability or interpretability in our project: the data scientists themselves needed to understand, scrutinize, and improve the model before they could confidently ship the tool to such a safety-critical domain. Unlike the physicians, they had to closely attend to all details of the model and ask a broader range of questions, not just *why* a particular prediction, but also *how* the model weighed different features, *what if* a feature changed, and so on. They were able to get some of these explanations through programming and visualization tools in the existing data science toolbox. Using this example, it is important to recognize that model explanations are sought by different types of *explanation consumers* who develop, use, manage, regulate, or are affected by AI systems. They may have different needs both in terms of the content of explanations and the presentations. It is therefore imperative to follow a user-centered approach to develop and design explainable AI systems.

example: A large project in a Danish hospital captured detailed minute-by-minute information about the medical secretaries who worked there. The goal was to provide an "objective" view of where their work happens in order to inform the plans for a new hospital building. By presenting the work back to the secretaries themselves, we (Gina and her team) were able to correct some of our key assumptions. The medical secretaries helped us understand how politicized the results were in their workplace because the original results suggested that they were no longer needed in the new hospital building and could work farther away from the clinical team they supported. By working with them, we were able to show how and why the project should include space for the medical secretaries, even though the purely quantitative data science results could have been narrowly interpreted to lead to a configuration that would have been less effective for them and the people they were supporting (Møller et al. 2020).

Similarly, Michael was asked to analyze the work of telephone operators at a US telecommunications company. For part of his analysis, he made video recordings of operators as they handled customer calls. With the permission of the operators' labor union, he set up the camera, and then he asked each video-recorded operator to watch their own video with him and provide their perspective about what had gone right and wrong on each call. After further review with the union, he was then able to present the operators' expert analyses of their work as a counterweight to technologists' ideas about the so-called simplicity of the operators' work.

A large part of data science is figuring out these goals. But working with groups has its own challenges (as we saw in chapter 7). It takes time to translate the results into language and stories that people can understand. It takes time to understand a community's concerns. Communities are not monolithic or homogeneous, and it takes work to tease out and adjudicate among people's different concerns. Working with communities means being led by other people's concerns, not our own (Hayes 2011). But it also means that communities become partners because they are experts in their experience.

Working with Journalists and the Media

News journalism has also taken a computational turn. *Data journalism* merges journalistic practices with data analysis and visualization. Journalists conduct research, gather information and data, analyze it, and report it. Journalists work with data science teams to present compelling, newsworthy information to the public. An example is the multinational collaborative effort of the Panama Papers, which exposed the widespread practice of offshore money laundering that led to at least 150 official inquiries, audits, or investigations in seventy-nine countries and resignations or removals from office of high-ranking officials, including the prime ministers of Iceland and Pakistan (Cabra 2017). At over 260 gigabytes of data, it was, at the time, the largest cache of leaked documents in the history of journalism, and the data team of the International Consortium of Investigative Journalists helped to make sense of these data for newspaper readers around the world.

Journalists also cover stories about data science projects, and they too tell stories about these projects. Meredith Broussard (2018) calls this "algorithmic accountability reporting," such as when the news organization ProPublica's 2016 "Machine Bias" story found that the algorithm used in judicial sentencing was biased against African Americans.

When we work with journalists on our data science project, it means revisiting the question "What story can I honestly and responsibly tell with this data for the public?" and working to create shared understanding.

Storytelling Strategies for Data Science

Below we summarize key storytelling strategies for data science to help projects consider people in their communication.

1. Telling stories takes time but is part of what a good data scientist should do, and it will help you communicate your results more effectively and memorably.

2. Consider your audience. Audiences matter. Differences across audiences mean changing how the story of your project is told. Keep in mind that you may have different audiences at different times. Just like you tell the story of your personal life differently to friends, parents, or work colleagues, you should remain focused on your audience and their needs.

3. Remember that good stories have narrative arcs. Telling the beginning, middle, and end of stories and anticipating what audiences want to hear next helps you connect with the people who use your data science.

4. Consider what the audience wants to do with your project results. The goals matter for data science. There is not one single way to tell the results of a project. Emphasize what helps people accomplish their goals while staying authentic to and transparent about the results.

5. Remember that you speak on behalf of your data. It is the responsibility of people communicating data science to convey it with accuracy but also emotion: uncertainty, urgency, persuasion.

6. Stories have an emotional hook. Interesting stories cause people to change their worldviews. Neuroscientists have shown that people make decisions through emotion (Damasio 2012). Decisions have an emotional element, and over and over again stories have been used as a catalyst for learning, change, and decision making.

7. Be aware of the fundamentals of how organizations and companies shape what and how and which stories can be told.

8. Be transparent with yourself and others as you craft your stories. Transparency is not just a function of algorithmic design but is a practice of the highest professional integrity for data science.

9. Experiment with your stories before you tell them to an audience. Earlier, we asked you to do an exercise of making your stories shorter and shorter to discover which parts were absolutely essential. Stories don't have to be as short as will fit into a twenty-three-second elevator pitch. But consider: We illustrated our points in this chapter with stories, and all of those stories were shorter than 125 words.

10. The "end" of the data science cycle is the beginning of conversations. Find the answers, but also help people figure out what issue is being addressed and how to ask the next questions.

11. Think about the end of the stories and what could go wrong with the data. What might be the unintended consequences? These can be surfaced by relying on speculative scenarios, considering design fictions, or crafting your own. What could be done to mitigate these unintended consequences?

Conclusion

Storytelling with data entails more than visualizing or presenting numerical results; you should consider who the audience is for your project, what they need, and how organizations shape how stories are told. Anticipating and planning for your audience's needs will help you better communicate your data science project. Experimentation, transparency, and flexibility are keys to success when using storytelling as a communication media to convey your data science outcomes.

People need stories, and much is gained by using stories in data science communication. Humans are hardwired to hear stories. Stories are compelling to us because they set up our interest in the things happening around us. Visualizations can serve as a great medium for storytelling, too. In this chapter, we looked at story structures and how visualizations fit into the storytelling path.

Thinking about stories as a mechanism for communication reminds us of the needs and strategies to consider when working with different groups. We must attend to different communities that have their own needs, values, and concerns. We work with experts in their own domains who might have different needs and concerns for communicating outcomes. Business leaders, policymakers, and journalists are types of data science audiences whose values are distinct. For journalists, the results of a data science project on policy may need to be communicated to a broader, less technical public audience in a transparent way. For policymakers, understanding the implications of a data science project on different members of society is tantamount. Business leaders and policymakers may also have a power relationship with the data scientists they work with. They may have their own values and needs that may constrain a shared project's objectives or agenda.

Recommended Reading

Knaflic, Cole Nussbaumer. 2015. *Storytelling with Data: A Data Visualization Guide for Business Professionals*. Hoboken, NJ: Wiley. This readable book, aimed at business professionals, presents visualization fundamentals in an accessible and engaging manner.

9

The Future of Human-Centered Data Science

Our goal for this book was to present a practical, real-world guide for doing data science about people, with people, and for people in a way that is both ethical and responsible. Throughout this book, we asked you to reflect on the choices and the responsibilities that people doing data science have. We hope that you have gained insight on how to think critically and reflexively in each step of the data science cycle.

We started this book with a definition of human-centered data science that recognizes that people are involved at every stage of the data cycle. This outlook takes seriously the audiences for data science and the communities that might be affected, sees data as a part of companies and systems that can shape the values of our projects, and holds central the value that much of our data represents real people and that our work has the potential for real consequences for them.

In this conclusion, we review the five key themes that cut across the previous chapters and examine human-centered data science as (1) ethical responsibility, (2) looking in the right places, (3) a collective practice, (4) communication, and (5) action. We then look toward a future when human-centered practices are more widely adopted in data science.

Human-Centered Data Science as Ethical Responsibility

Data science systems can provide powerful insights that have a significant influence on the world. We have tried to make clear that the practice of data science is not data-driven alone; it is also powerfully shaped by human assumptions, human beliefs, human values, and human actions. We hope that the stories we have told have conveyed our sense of humility: there is simply no easy checklist for doing data science ethically and responsibly. Data science bestows power on people who design, create, and operate data science systems. What is your responsibility to yourself and to others when you work in data science? How will you respond to the responsibility that goes along with the power of your system to affect people's lives?

We have tried to show that such questions are at the heart of human-centered data science. We described several different approaches to ethical research and how these traditions can inform the design of data science projects. We reviewed ethical issues that arise in cleaning and designing a dataset and in training a model, and we described many possible sources of bias that could arise at different stages.

Decisions made at each stage of the data science cycle require ethical attention to doing the work responsibly and treating with ethical care the people potentially affected by our data, models, and results. *Data discovery* is a moment to raise questions about what is relevant in terms of data and datasets. *Data capture* entails choices in creating or assembling a dataset to be used in a data science project. *Data curation* requires excluding some records and including others—choices that have implications for people's decisions about what counts as an outlier and what counts as centrally important to a project, potentially flattening the complexity necessary for fully parsing human behavior. *Data design* describes the process of categorizing variables using terms like "low," "medium," and "high" or creating new variables through feature engineering or other combinations of existing variables. *Data creation* describes what happens when people make decisions about the ground truth of a variable. Each step relies on assumptions, in the best case, or biases, in the worst. The methods and tools to detect and mitigate biases will continue to evolve in data science, and ethical, reflexive practice will continue to be central to doing good work in human-centered data science.

Throughout the book we also discussed how large-scale data presents a series of challenges to people's rights to privacy. We showed how seemingly anonymous and innocuous data can quickly become identifiable and potentially harmful. We have ethical responsibilities to the people who might be harmed by our projects, and we should ensure that we keep these responsibilities in sight. We showed how reflecting on your data science practices makes for better *and* more ethical data science.

Human-Centered Data Science as Looking in the Right Places

Human-centered data science is about framing the right questions and looking in the right places for their answers. This process occurs throughout the data science cycle.

Good data science projects are anchored with good questions and are based on having goals for the project—what you want to know—not simply data that happens to be on hand. Anchoring your project with good questions means that you will have to do more work and not look for the answers in the most convenient place. But the result is that you have a better idea of what data you need to answer your question and how much of that data you need to answer it convincingly.

Social media data, especially from Twitter, has been one such "easy" place for data science to look. However, certain problems emerge when hashtag and keyword analyses "select on the dependent variable," or the self-selection to include these terms or hashtags in a post, as we discussed in chapter 6. Also, people use social media conversationally: irony, humor, and sarcasm may be clear from context but missed in large-scale data.

There are other reasons that human-centered data science puts good questions at the heart of our projects. Asking questions first helps us think through the ethics of formulating our project and helps prime us for what we *should* be looking for. Developing interesting questions first helps us know where to look and clarifies what we hope to discover when we do. Good, clear questions help us think about the problem or puzzle at the center of our projects and how we might better communicate the story of that puzzle to others.

Clear questions help us responsibly and ethically use the amount of data proportionate for our needs: neither too much data to answer our questions, unnecessarily exposing people's information to the risks of reidentification, nor too little to responsibly answer our questions (UK Government Digital Service 2020).

Starting with our questions helps us pay attention to the choices we make across the data science cycle. Keeping a focus on our questions and our initial objectives or goals helps us figure out what to do at those moments. Reflecting on this process—and our active work of seeking the answers—helps us to clarify our choices. There are many of these choices in data science: the number of clusters for an algorithm to find, the technical parameters we use to make a model "work," or the statistical ranges or cutoffs for our models. Rather than being data-driven, these decisions are always a combination of human insight and algorithmic support for that insight. Being able to articulate the question we are asking with our project helps us justify these choices.

Starting with our questions also demands that we reflect and document these choices so that we can show why we thought the place that we were shining our light is the "right" one. This book has pushed you to think about data science as a set of practices with multiple possibilities and not simply a right or wrong way to ask questions. Being able to reflect on this work as we make the many choices required for a data science project helps us remember that we are responsible and accountable for how our models and pipelines come to their final shape, and what questions we might want to revisit, starting the cycle again. It also helps others understand what stories our data tell and why.

Human-Centered Data Science as a Collective Practice

The practices that we describe in this book see data science as a collaborative process of different stakeholders, often working with people's data. Throughout this book, we have provided reflective questions as the starting point for discussion or introspection about your practices and how you collaborate with others. Throughout this book we have urged you to think about and anticipate how others might use your data science work. We gave pointers on many different human-centered practices and methods through which you could share knowledge with others. We described methods and practices that you can use in various relationships and at various scales, from simple one-on-one interviews to work with larger communities and organizations.

We believe that working with others will help you to choose the right places to look for your insights. Everyone can identify problems, and everyone can generate solutions. When we think together, we are more powerful than when we think alone. That can mean thinking across business functions in a company, thinking across disciplines in academic research, or thinking across different types of stakeholder knowledge. In several chapters we discussed how others can inform your data science project from their own knowledge—insight that you may not have or may not yet realize how it can be useful to you. Learning to empower others with data science is one of the goals we have set forth in this book. Bringing humility to these partnerships and collaborations is what having a human-centered approach implies.

Data scientists rarely work alone, although many are trained individually. Learning good teamwork takes time and practice. Data science teams often include members with different roles, training, and backgrounds. Learning how to work across these differences is a skill that will serve you well throughout your career. Good data scientists learn to negotiate issues across a variety of teams and styles as part of doing data science.

Collaborators have different ways of working, and we have pointed out many issues to consider when working with others. Are you thinking about the various roles people play on data science teams? Are you working with others to empower them to ask questions of the data? Are you anticipating how your work could expose others to risk or harm? There is no clear checklist for how to work well with others. But across these chapters we have pushed you to reflect on data science as a collective, not individual, practice.

Human-Centered Data Science as Communication

Our approach to data science sees the output of models as only one step in a larger and richer process that, at its heart, is about using data to have an impact on a real-world problem. For data science to "work" this way, communication is crucial. How we and others present and represent our work is central to ethical, responsible, and human-centered data science.

In chapter 8, we talked about data science as storytelling. Our goal was to encourage you to think about how to tell the story of your projects or results; the storytelling technique helps us to keep in mind the people in the various audiences for our work. Learning to craft a good story helps us frame our presentations for people so that they ask themselves, and us, "What happened next?" Stories help us remember that communication is a process that involves others, not just ourselves and our intentions. The results of our projects are not the end of our job, but in some ways only the beginning. Telling stories with our data takes time but is part of what a good data scientist should do.

Many projects rely on good communication across a data science team and with our partners and collaborators. Learning to bridge the differences in values—whether these are disciplinary or knowledge differences, differences in personal experiences and histories, or differences in power—is a valuable professional skill, and one that, as we learned from the head of data science at the *New York Times*, is in high demand among companies that employ data scientists.

Communication also plays a role in other parts of the data science cycle. Learning to ask the right questions about the provenance of a dataset prevents later errors. Learning that others may want us to communicate this information after our part in the project is over teaches us that documentation is important. Documentation is where we can reflect on the choices that we made throughout the data science cycle and where we can share those choices transparently and responsibly. In other words, documentation is another area for communication in human-centered data science.

Human-Centered Data Science as Action

In her landmark poem "Sources," Adrienne Rich (1986) asks repeatedly, "With whom do you believe your lot is cast?" While we would like to think that our data science systems serve all of humanity, Value Sensitive Design helps us to see that systems often have different impacts for people in different organizational or societal positions. Our systems can favor certain groups or social classes over others. Our systems can create haves and have-nots. Our systems can magnify privilege and intensify marginalization. To put it bluntly, our systems can produce winners and losers and can severely impact both direct and indirect stakeholders. We hope that, as you design and build your data science system, you will pause and think about the broader world into which you will insert your system. Who will benefit from your system? Who will be harmed? With whom do you believe your lot is cast?

When we make a system that will influence other humans, we engage in a kind of world-making (Floyd et al. 1992). Data science systems can have life-altering influences, from early diagnosis of cancer to prison-sentencing recommendations that can lock people away from their families for decades. When we think about data science systems, we are already taking a step toward creating future worlds that we and our communities will inhabit (Muller and Erickson 2018). We hope that you will use human-centered methods and perspectives to carefully consider the kinds of worlds that you are designing and the consequences to yourself and others of living in those worlds.

When you have answered the question "With whom do you believe your lot is cast?" then we hope that you will work with other people who know different things from what you know. Collectively, we can choose the world or worlds that we want to live in *together*. That kind of mutual education is one of the greatest joys of human-centered data science.

Looking Forward in Human-Centered Data Science

Where might human-centered data science go? We wrote this book because we think that data science will be better with a human-centered approach. We have values about what we think "better" data science looks like. Hopefully, by this conclusion you also see that these very human values can inform data science practices that respond to the communities we may impact; in so doing, we share results with audiences in transparent and responsible ways, consider the people whose data we work with to try to reduce the harm that our projects might cause them, design and communicate clear goals for our projects, and work well with our teams, collaborators, and community partners. We think this is the way forward for data science that has a positive impact on improving the world.

Such a future is not a given. Data is power, as we learn from the many community efforts, from Data for Black Lives to Code for America to Community Data Science Workshops.

This book is primarily about people: the people whose data we analyze, the people we collaborate with, the people who are affected by the work we do, and also data scientists themselves. As we reviewed in our discussion of Value Sensitive Design in chapter 7, people are also the users of data science systems—the direct stakeholders—and people are also the colleagues, friends, family members, and associates of the users—the indirect stakeholders. How will you communicate with others about your plans, designs, and intentions for your data science system? What is the story of your system? The social impact of a data science system can be broad and profound. We hope that you will take to heart the phrase "these are people" when you consider how your system and methods will affect others. And we hope that you join us in shaping this future for data science.

Glossary

Action research is a social science method that seeks transformative change through the simultaneous process of taking action and doing research, which are linked together by critical reflection.

Adversarial data collection means that data scientists will deeply interrogate what types of data they need for their analysis, as opposed to using everything that is available just because it is there.

Adverse impact is the negative effect an unfair and biased selection procedure has on a class of people.

Affordances are an environment's properties that show the possible actions users can take within it, suggesting how users may interact within that environment.

Algorithmic bias is the inherent bias of algorithm designers encoded in decisions made while crafting an algorithm.

Anderson–Darling test is a statistical test of whether a given sample of data is drawn from a given probability distribution. In its basic form, the test assumes that there are no parameters to be estimated in the distribution being tested, in which case the test and its set of critical values are distribution-free.

Annotation/annotating or **labeling** typically takes a set of unlabeled data and augments each piece of that data with meaningful tags that are informative.

ANOVA (analysis of variance) is a parametric statistical model that can be used to compare means of two or more samples.

Appropriation occurs when an artifact is designed for one purpose but people use it for a different purpose.

Arithmetic mean is the average of a set of numerical values, calculated by adding them together and dividing by the number of terms in the set.

Artificial neural networks (ANNs), or just neural networks, are a type of modeling approach inspired by a simplified model of how the human brain operates in terms of developing a network of "neurons" and "edges" that connect them.

Bayesian hierarchical modeling is a statistical model written in multiple levels (hierarchical form) that updates the probability of parameter values using the Bayesian method. The intuition behind this kind of modeling is that oftentimes, in dealing with complex data and data science problems, we are forced to recognize that entities exist in nested and intersectional groups with each other. Thus, when new data or evidence comes in, probabilities should get updated conditioned on these hierarchical group structures.

Bayesian networks are a kind of statistical model that represents a set of variables and their conditional dependencies via nodes and edges in a graph. Bayesian networks are ideal for taking an event that occurred and predicting the likelihood that any one of several possible known causes was the contributing factor.

Bayesian statistical models are a popular family of parametric statistical models. They assume that there are no hypothetically "true" parameter values, such as mean or variance, for a given population, but that parameter values keep changing from dataset to dataset.

Bias is a disproportionate weight in favor of or against an idea or thing, usually in a way that is unfair.

Binary data is data whose unit can take on only two possible states, traditionally labeled as 0 and 1.

Capture refers to when someone finds data and brings it into a dataset for analysis.

Classification accuracy is the fraction of correct predictions made by the classification algorithm out of the total number of predictions made.

Cluster analysis is a modeling approach that groups a set of objects in such a way that objects in the same cluster are more similar (in some way) to each other than objects in another cluster.

Cognitive walkthrough is a usability evaluation method in which one or more evaluators work through a series of tasks and ask a set of questions from the perspective of the user. The focus of the cognitive walkthrough is on understanding the system's learnability for new or infrequent users.

Computational social science is the field of social science that uses computational approaches in studying social phenomena.

Confusion matrix is a table that is often used to describe the performance of a classification model on a set of test data for which the true values are known. It allows the visualization of the performance of an algorithm.

Construct validity refers to the appropriateness, accuracy, and quality of labels applied to data based on observation or measurement, specifically whether and how well a label reflects the intended construct.

Context is information about people, situations, and environments that are outside the data but associated with the data or its generation in some way.

Context-rich data refers to datasets that provide enough information about human activity, especially social interactions and cultures, for the data scientist to account for the various differences present in human behavior.

Critical data studies is the systematic study of data and its role in supporting power within society.

Critical race approaches look at how data might intersect with racial inequalities in society.

Curation occurs when someone makes a collection of items (like data records) and organizes the items according to a scheme or a plan.

Curb cut is the name for a ramp in a physical curb that allows wheeled travel from curb to street and from street to curb. People with many different types of needs benefit from curb cuts. See also **electronic curb cut**.

Data acquisition is the activity of finding/getting/creating data for data science work.

Data encoding means classifying each variable into a type and then determining which visual attributes represent these data types most effectively.

Data feminism is a way of thinking about data science and data ethics that is informed by the ideas of feminism.

Data science ethnography is using the tools of ethnography to study data science in action, focusing on how people "work with large and complex datasets, and the institutions, programs, and communities" that support data science.

Data wrangling refers to the often messy work of getting data ready for analysis. It involves cleaning the data, which includes deciding how to deal with missing values and outliers, and tidying the data into a more appropriate format.

Decision tree is a decision-support tool that uses a treelike model of decisions and their possible consequences, including chance event outcomes, resource costs, and utility.

Design fictions are pieces of speculative storytelling that use fiction to enable deeper exploration of the potential implications of design elements.

Design justice is an approach that seeks to use design to rectify injustices or inequalities.

Differential privacy is a system for publicly sharing information about a dataset by describing the patterns of groups within the dataset while withholding information about individuals in the dataset.

Digital ethnography describes the process and methodology of doing ethnographic research in a digital space.

Dimension(ality) reduction is the process of reducing the number of random variables under consideration in a statistical model by obtaining a set of underlying, latent variables.

Discovery occurs when a data science worker finds a dataset that is ready to be used, without having to do much extra work to begin to analyze it.

Domain experts, or **subject-matter experts**, are people who have in-depth knowledge outside of data science. Their expertise in the particular situation, context, data, or problem at hand helps in the design of better human-centered data science projects.

Dummy variable is a variable that takes only the value 0 or 1 to indicate the absence or presence of some categorical effect that may be expected to shift the outcome.

Electronic curb cut allows people with disabilities to use electronic technologies more easily. Electronic curb cuts include screen readers (text to speech) and voice interfaces (speech to text), as well as closed captioning and various haptic (touch-based) user interfaces. Like physical **curb cuts**, electronic curb cuts provide benefits to a wide range of users, including but not limited to people with disabilities.

Ethnography is a technique rooted in the fields of anthropology and sociology that involves the in-depth examination of participants in a particular culture or social situation to not only reveal specifics of their behavior, but also explore the underlying, perhaps unspoken, reasons for their behavior as individuals and groups.

Experimental design refers to the process of formal analysis and design of experiments in order to conduct statistical modeling—be it nonparametric or frequentist or Bayesian in nature. It is the design of any task that aims to describe and explain the variation of information under conditions that are hypothesized to reflect the variation.

Explainability is the extent to which the internal mechanics of a machine learning system can be explained in human terms.

F1 score is the harmonic mean of **precision** and **recall**.

False negative is an error in which a test result improperly indicates no presence of a condition (the result is negative), when in reality it is present. Please see True negative for comparison.

False positive is an error in data reporting in which a test result improperly indicates the presence of a condition, such as a disease (the result is positive), when in reality it is not present. Please see True positive for comparison.

Feature is an individual measurable property or characteristic of a phenomenon being observed.

Feature engineering is the process of using domain knowledge to extract features from raw data using data mining techniques.

Feature extraction starts from an initial set of measured data and builds derived values (features) intended to be informative and nonredundant, facilitating the subsequent learning and generalization steps and in some cases leading to better human interpretations.

Feminist approaches constitute an array of theories and methods that focus on the personhood of all people. Feminist approaches balance understanding all persons as equal and, simultaneously, understanding each person as having both unique personal needs and aspirations and group-based needs and aspirations. Feminist approaches consider power relations and tend to reject norms (especially compulsory norms) and "universals" in favor of a focus on margins and marginalized peoples.

Focus groups are small but demographically diverse groups of people whose reactions are studied, especially in market research or political analysis, in guided or open discussions about a new product or something else to determine the reactions that can be expected from a larger population.

Frequentist statistical models are the most widely used family of parametric statistical models. Here statistical inference draws conclusions from sample data by emphasizing the frequency or proportion of the data.

Generative models that do semi-supervised learning can be viewed as an extension of supervised learning (classification plus information about some probabilities of belonging to a given class) or as an extension of unsupervised learning (clustering plus some labels).

Geometric mean indicates the central tendency or typical value of a set of numbers by using the product of their values. It is defined as the nth root of the product of n numbers.

Ghost work refers to the invisible labor that powers technology platforms.

Git is a distributed version-control system for tracking changes in source code during software development. It is designed for coordinating work among programmers, but it can be used to track changes in any set of files.

GitHub is a Git repository hosting service. It provides a web-based graphical interface, access control, and several collaboration features, such as wikis and basic task management tools for every project.

Graph-based models develop a graph representation of data where each node is either a labeled or unlabeled example. Graph similarity methods like centrality metrics or other distance metrics are then used with the intuition that "closer" labeled and unlabeled pairs probably belong to the same category and hence can be classified into the same class.

Grounded theory is a systematic methodology in the social sciences involving the construction of theories through methodical gathering and analysis of data using abductive reasoning and well-defined procedures and protocols, such as theoretical sampling and the constant comparison of data with data and of data with theory. See also **thematic analysis.**

Ground truth refers to information provided by direct observation as opposed to information provided by inference.

Heuristic evaluation is a usability evaluation method that helps to identify usability problems in the user interface design. It involves evaluators examining the interface and judging its compliance with recognized usability principles.

Human-centered algorithm design is a framework for designing algorithms that aims to predictively model human behavior using theoretical, contextual, participatory, and speculative design approaches.

Human-centered data science includes a focus on data about people that appreciates the unique threats to autonomy, privacy, and ethics that large-scale data presents. It also refers to combining human-centered design approaches to data science. Human-centered data science also encompasses attention to how people use the results of data science and the role of data science practitioners in making their work intelligible and understandable to different audiences. It also refers to the values and concerns of organizations that might shape or direct data science projects.

Human-centered design is an approach to problem solving, commonly used in design and management frameworks, that develops solutions to problems by involving the human perspective in all steps of the process. Human involvement typically takes place in observing the problem within context, brainstorming, conceptualizing, developing, and implementing the solution.

Information visualization refers to the visual representation of abstract information that may not have a physical reference, such as sales or stock prices.

Internal validity is the extent to which a piece of evidence supports a claim about cause and effect, within the context of a particular study.

Internet of things (IoT) describes the network of physical objects—"things"—that are embedded with sensors, software, and other technologies for the purpose of connecting and exchanging data with other devices and systems over the internet.

Interpretability of a model means that someone can interpret how the model delivers its particular results and the extent to which a cause and effect can be observed within a model.

Interpretable machine learning focuses on the development or study of statistical or automated techniques that are understandable or transparent to people. (See also **interpretability**.)

Interpretation is a key component of the data science cycle.

Intersectionality is a theoretical framework for understanding how aspects of one's social and political identities (race, class, gender) might combine to create unique modes of discrimination.

Interviews in qualitative research are conversations in which questions are asked to elicit information. The interviewer is usually a professional or paid researcher, sometimes trained, who poses questions to the interviewee in an alternating series of usually brief questions and answers.

Jupyter notebook is a web-based interactive computational environment for creating code-based documents that can also include comments and visualizations.

K-anonymity is a property of a dataset, usually used to describe the level of anonymity in a dataset. A dataset is k-anonymous if every combination of identity-revealing characteristics occurs in at least k different rows of the dataset.

Kolmogorov–Smirnov test is a nonparametric test of the equality of continuous, one-dimensional probability distributions that can be used to compare a sample with a reference probability distribution (one-sample K–S test) or to compare two samples (two-sample K–S test).

Kruskal–Wallis test is a nonparametric method for testing whether samples originate from the same distribution. It is used for comparing two or more independent samples of equal or different sample sizes.

Labeled data is a group of samples that have been tagged with one or more explanatory labels.

Latent Dirichlet allocation is a generative statistical model that allows sets of observations to be explained by unobserved groups (topics) that explain why some parts of the data are similar.

Linear regression is an approach that maps the relationship between a response or dependent variable and one or more (usually more) predictor or independent variables. The objective is to produce the best-fit straight line that produces the most average outcome.

Logistic regression models are used to calculate the probability of a particular event occurring (e.g., pass/fail) given a set of predictor or independent variables.

Measurement plan is a set of definitions and plans for action that govern a data science project in many ways. Measurement plans define the relevant "data" for a project and indicate some of the project's goals. Measurement plans may be well-documented and recorded, or they may be transient and difficult to retrieve.

Median is a value lying at the midpoint of a frequency distribution of observed values, such that there is an equal probability of falling above or below it.

Mixed methods refers to the systematic integration, or "mixing," of multiple research methods, often quantitative and qualitative, within a single project to take advantage of the strengths of both.

Model refers to the core statistical engine that computes a predictive relationship in supervised machine learning. See also **pipeline**.

Multiclass data is data whose unit can take on three or more possible states.

Natural language processing (NLP) is a subfield of linguistics, computer science, and machine learning that is concerned with allowing computer systems to process natural language—how people naturally speak. In practice, that often means building algorithmic tools that can process (translate, get semantic structure for, categorize) large amounts of textual or voice data.

Network science is an academic field and a research methodology that studies complex networks such as telecommunication networks, computer networks, biological networks, cognitive and semantic networks, and social networks, considering distinct elements or actors represented by nodes (or vertices) and the connections between the elements or actors as links (or edges).

Nonparametric statistics are a family of statistical models that make no assumptions about parameters in the underlying data (common examples of parameters are the mean and variance). Nonparametric statistics is either distribution-free or assumes a specified distribution but with the distribution's parameters unspecified.

Normalized is a word used in at least two ways in data science. (1) It may refer to a normal distribution. For some data science methods such as **ANOVA, principal component analysis**, and regression, the statistical basis of the methods assumes a normal distribution as defined in statistics. (2) Normalization is also used to describe converting a variable into a proportional measure. For example, we might want to know whether children are considered by their peers to be "tall." However, a tall four-year-old is shorter than a short eight-year-old. We could normalize the children's height within each age range by finding the tallest child of that age and then dividing the height of each child in that age range by the tallest child's height. The tallest child in each age range would now have a normalized height of 1.0, and the other children would have proportional normalized heights less than 1.0.

Observation is a qualitative social science method where researchers watch and note how people do particular things while producing careful and systematic notes, known as field notes.

Operationalize is the process of defining the measurement of a phenomenon that is not directly measurable, though its existence is inferred by other phenomena. This is used in social sciences to measure theoretical constructs (e.g., depression) that usually cannot be measured directly.

Ordinal is a categorical, statistical data type where the variables have natural, ordered categories and the distances between the categories are not known.

Outliers are (typically quantitative) data values that seem to be *very* different from other data values in the same variable. True outliers can distort an analysis. However, outliers may also reflect previously unacknowledged classes or subclasses within a data sample.

Participatory analysis is a qualitative social science method that invites people to construct a description of their experience and then interpret their own description; both description and interpretation become data for analysis.

Participatory design is a large body of methods and theories in which the "users" of technologies take the role of active designers or co-designers. Related methods involve "users" as co-analysts or co-evaluators. Participatory design originated in the Scandinavian workplace democracy movement and continues to promote the political and workplace power of the users of technologies.

P-hacking is the misuse of data analysis to find patterns in data that can be presented as statistically significant, thus dramatically increasing and understating the risk of false positives.

Pipeline is a series of steps in a data science cycle, beginning with data and ending with accuracy measurements, possibly bias mitigation, and deployment for actual usage. See also **model**.

Pragmatic tradition is an approach to social science that is a committed to end-causes and outcomes of practice (such as social change) and interventions that transform society.

Precision is the fraction of relevant instances among the retrieved instances.

Principal component analysis (PCA) is a way to reduce complexity of datasets from many, possibly correlated variables to a few, uncorrelated variables by doing some data transformations (usually orthogonal).

Principle of Consistency means that the properties of the visual attributes should match the properties of the data variables. That is, one-dimensional data such as cost should not be represented by area or volume.

Principle of Effectiveness suggests that a visualization is more effective than another visualization if its information is more readily perceived than the information in the other.

Principle of Expressiveness suggests that the visual attributes should express all the facts in the set of data and only the facts in the data.

Principle of Importance Ordering means that we should encode the most important information in the most effective way. Since humans can detect position most accurately, use it to encode your most important quantitative data.

Provenance is the history of an artifact, such as a dataset or a pipeline. The provenance tells you the source of the dataset, who worked with the dataset, and how they may have modified it.

Qualitative data analysis entails ways of analyzing non-numerical data. **Grounded theory** and **thematic analysis** are two qualitative research methods.

Qualitative research methods are social science methods focused on *qualities*: meanings and values and processes rather than easily measured amounts. Qualitative research aims to capture how people make sense of their own situation and reveal the context where they do so. Qualitative research focuses on how people experience and give meaning to the things that happen to them. It often relies on the researcher's interpretation.

Quantitative methods refer to analyzing numerical data using mathematical or statistical approaches. In the social sciences, the term quantitative methods groups together many different sources of data (surveys, documents) to focus on the data type (numerical) and the techniques of analysis (primarily statistical).

Random forests are a machine learning method for classification, regression, and other tasks that operate by constructing a multitude of decision trees at training time and outputting the class that is the mode of the classes (classification) or mean prediction (regression) of the individual trees.

Recall is the fraction of the total amount of relevant instances that were actually retrieved.

Reflexivity is the process by which the researcher reflects on the data collection, analysis, and interpretation process.

Regression is a statistical model that estimates a measure of the relation between independent (or input) variables and a dependent (or output) variable.

Reification is a complex idea for when you treat something immaterial—like happiness, fear, or evil—as a material thing.

Reliability is the overall consistency of a measure. A measure is considered to have a high reliability if it produces similar results under consistent conditions.

Reproducibility is a word used in at least two ways in data science. (1) It may refer to designing your projects so that others can test whether they would reach the same results. (2) **Reproducibility** is also used to describe the provision of enough detail about study procedures and data so that the same procedures could be exactly repeated.

Science and technology studies is the study of how society, politics, and culture affect scientific research and technological innovation and how these, in turn, affect society, politics, and culture.

Scientific visualization usually involves displaying data that has a physical position in space, such as a gas combustion dataset.

Self-organizing maps are a type of **artificial neural network** (ANN) that are trained using unsupervised learning to produce a low-dimensional representation of the input data, called a map. It is a method for dimensionality reduction.

Semi-supervised machine learning is an approach to machine learning that combines a small amount of labeled data with a large amount of unlabeled data during training.

Skewed data is a term to describe when a curve appears distorted or shifted either to the left or to the right, with one tail longer than the other, in a statistical distribution.

Social data science is a flavor of the data science discipline where the data relates to individual and social behavior; social science with generation and analysis of real-time transactional data is at its center.

Standpoint is a position from which objects or principles are viewed and according to which they are compared and judged. The inequalities of different social groups create differences in their standpoints.

Statistical hypothesis testing is a method of statistical inference. It sets up an experiment where a hypothesis that is testable on the basis of observing a process is modeled using a set of random variables in order to make a quantitative decision about rejecting or accepting the hypothesis.

Statistics is the discipline that concerns the collection, organization, analysis, interpretation, and presentation of data.

Supervised machine learning is a large family of data science modeling methods in which we want to predict a variable (called **ground truth**) on the basis of other variables.

Support vector machines (SVMs) are a type of **supervised machine learning** model that aims to construct an optimum straight line that separates sets of points that are labeled as belonging to different categories.

Thematic analysis is a principled six-step method for analyzing (usually) qualitative data. See also **grounded theory**.

Tidy data is a standard way of mapping the meaning of a dataset to its structure. A dataset is **messy** or tidy depending on how rows, columns, and tables are matched up with observations, variables, and types.

Topic modeling is a data science modeling approach to analyze large text corpora for overarching "topics" or distributions of words that recur with statistical regularity.

Training dataset is an initial set of data used to train an algorithm. Training data is a certain percentage of an overall dataset along with a testing set.

Transparency is the principle suggesting that the factors that influence the decisions made by algorithms should be visible, or transparent, to the people who use, regulate, and are affected by systems that employ those algorithms.

True negative is when a test result indicates no presence of a condition (the result is negative) and in reality the condition is not present. See False negative for comparison.

True positive is when a test result indicates the presence of a condition (the result is positive) and the condition is indeed present. See False positive for comparison.

Unsupervised machine learning is a type of machine learning that looks for previously undetected patterns in a dataset with no preexisting labels and with minimal human supervision.

Validity is the extent to which a concept, conclusion, or measurement is well-founded and likely to correspond accurately to the real world. The validity of a measurement tool is the degree to which the tool measures what it claims to measure.

Value Sensitive Design (VSD) is a theoretically grounded approach to the design of technology that accounts for human values in a principled and comprehensive manner.

Values in design is a movement that goes beyond traditional requirements engineering to consider individual and social values as equally important inputs to the technology design process.

Vignettes are short stories in qualitative social science research, such as in-depth user profiles, that can be interpreted holistically or thematically.

Visual analytics is a combination of highly interactive visual interfaces and statistical learning algorithms.

Wilcoxon signed-rank test is a nonparametric statistical hypothesis test used to compare two related samples, matched samples, or repeated measurements on a single sample to assess whether their population mean ranks differ.

References

Alcoff, Linda. 1991. "The Problem of Speaking for Others." *Cultural Critique* 20: 5–32. https://doi.org/10.2307/1354221.

Allyn, Bobby. 2020. "IBM Abandons Facial Recognition Products, Condemns Racially Biased Surveillance." National Public Radio, June 9. https://www.npr.org/2020/06/09/873298837/ibm-abandons-facial-recognition-products-condemns-racially-biased-surveillance.

Alsheikh, Tamara, Jennifer A. Rode, and Siân E Lindley. 2011. "(Whose) Value-Sensitive Design: A Study of Long-Distance Relationships in an Arabic Cultural Context." In *Proceedings of the ACM 2011 Conference on Computer Supported Cooperative Work* (CSCW '11), 75–84. Hangzhou: Association for Computing Machinery. https://doi.org/10.1145/1958824.1958836.

Anderson, Michael R., and Michael Cafarella. 2016. "Input Selection for Fast Feature Engineering." In *2016 IEEE 32nd International Conference on Data Engineering (ICDE)*, 577–588. Helsinki, Finland. https://doi.org/10.1109/ICDE.2016.7498272.

Anscombe, Francis J. 1973. "Graphs in Statistical Analysis." *The American Statistician* 27 (1): 17–21.

Aragon, Cecilia, and Katie Davis. 2019. *Writers in the Secret Garden: Fanfiction, Youth, and New Forms of Mentoring.* Cambridge, MA: MIT Press.

Aragon, Cecilia, and Sarah S. Poon. 2007. "The Impact of Usability on Supernova Discovery." In *Workshop on Increasing the Impact of Usability Work in Software Development, CHI 2007: ACM Conference on Human Factors in Computing Systems: San Jose, CA.* New York: ACM Press.

Aragon, Cecilia, Clayton Hutto, Andy Echenique, Brittany Fiore-Gartland, Yun Huang, Jinyoung Kim, Gina Neff, Wanli Xing, and Joseph Bayer. 2016. "Developing a Research Agenda for Human-Centered Data Science." In *Proceedings of the 19th ACM Conference on Computer Supported Cooperative Work and Social Computing Companion* (CSCW '16 Companion), 529–535. New York: Association for Computing Machinery. https://doi.org/10.1145/2818052.2855518.

Aragon, Cecilia, Stephen J. Bailey, Sarah Poon, Karl Runge, and Rollin C. Thomas. 2008. "Sunfall: A Collaborative Visual Analytics System for Astrophysics." *Journal of Physics: Conference Series* 125 (July): 012091. https://doi.org/10.1088/1742-6596/125/1/012091.

Aragon, Cecilia R., and Alison Williams. 2011. "Collaborative Creativity: A Complex Systems Model with Distributed Affect." In *Proceedings of the SIGCHI Conference on Human Factors in Computing Systems* (CHI '11), 1875–1884. Vancouver: Association for Computing Machinery. https://doi.org/10.1145/1978942.1979214.

Aragon, Cecilia R., and David Bradburn Aragon. 2007. "A Fast Contour Descriptor Algorithm for Supernova Image Classification." *Proceedings of SPIE 6496, Real-Time Image Processing 2007*: 649607. International Society for Optics and Photonics. https://doi.org/10.1117/12.703666.

Aragon, Cecilia R., and Sarah Poon. 2011. "No Sense of Distance: Improving Cross-Cultural Communication with Context-Linked Software Tools." In *Proceedings of the 2011 iConference* (iConference '11), 159–165. Seattle: Association for Computing Machinery. https://doi.org/10.1145/1940761.1940783.

Aragon, Cecilia R., Sarah Poon, and Claudio T. Silva. 2009. "The Changing Face of Digital Science: New Practices in Scientific Collaborations." In *CHI '09 Extended Abstracts on Human Factors in Computing Systems* (CHI EA '09), 4819–4822. Boston: Association for Computing Machinery. https://doi.org/10.1145/1520340.1520749.

Aragon, Cecilia R., Sarah S. Poon, Andrés Monroy-Hernández, and Diana Aragon. 2009. "A Tale of Two Online Communities: Fostering Collaboration and Creativity in Scientists and Children." In *Proceedings of the Seventh ACM Conference on Creativity and Cognition* (C & C '09), 9–18. Berkeley: Association for Computing Machinery. https://doi.org/10.1145/1640233.1640239.

Aragon, Cecilia R., Sarah S. Poon, Gregory S. Aldering, Rollin C. Thomas, and Robert Quimby. 2008. "Using Visual Analytics to Maintain Situation Awareness in Astrophysics." In *2008 IEEE Symposium on Visual Analytics Science and Technology*, 27–34. Columbus, Ohio: Institute of Electrical and Electronics Engineers. https://doi.org/10.1109/VAST.2008.4677353.

Arayasirikul, Sean, Erin C. Wilson, and Henry F. Raymond. 2017. "Examining the Effects of Transphobic Discrimination and Race on HIV Risk among Transwomen in San Francisco." *AIDS and Behavior* 21 (9): 2628–2633.

Arnstein, Sherry R. 1969. "A Ladder of Citizen Participation." *Journal of the American Institute of Planners* 35 (4): 216–224. https://doi.org/10.1080/01944366908977225.

Asad, Mariam. 2019. "Prefigurative Design as a Method for Research Justice." *Proceedings of the ACM on Human–Computer Interaction* 3 (CSCW). https://doi.org/10.1145/3359302.

Asakawa, Chieko. 2005. "What's the Web Like If You Can't See It?" In *Proceedings of the 2005 International Cross-Disciplinary Workshop on Web Accessibility* (W4A), 1–8. Chiba, Japan: Association for Computing Machinery. https://doi.org/10.1145/1061811.1061813.

Babbie, Earl. 2017. *The Basics of Social Research*. 7th ed. Boston: Cengage Learning.

Baldassarre, Michele. 2016. "Think Big: Learning Contexts, Algorithms, and Data Science." *Research on Education and Media* 8 (2): 69–83.

Baumer, Eric P. S., David Mimno, Shion Guha, Emily Quan, and Geri K. Gay. 2017. "Comparing Grounded Theory and Topic Modeling: Extreme Divergence or Unlikely Convergence?" *Journal of the Association for Information Science and Technology* 68 (6): 1397–1410. https://doi.org/10.1002/asi.23786.

Baumhauer, Judith F. 2017. "Patient-Reported Outcomes—Are They Living Up to Their Potential?" *New England Journal of Medicine* 377 (July 6): 6–9. https://doi.org/10.1056/NEJMp1702978.

Bean, Jonathan, and Daniela Rosner. 2014. "Big Data, Diminished Design?" *Interactions* 21 (3): 18–19.

Becher, Tony. 1989. *Academic Tribes and Territories: Intellectual Enquiry and the Cultures of Disciplines*. Milton Keynes, UK: Society for Research into Higher Education and Open University Press.

Bellamy, Rachel K. E., Kuntal Dey, Michael Hind, Samuel C. Hoffman, Stephanie Houde, Kalapriya Kannan, Pranay Lohia, Jacquelyn Martino, Sameep Mehta, Aleksandra Mojsilović et al. 2019. "AI Fairness 360: An Extensible Toolkit for Detecting and Mitigating Algorithmic Bias." *IBM Journal of Research and Development* 63 (4/5): 1–15. https://doi.org/10.1147/JRD.2019.2942287.

Benjamin, Ruha. 2019. *Race after Technology: Abolitionist Tools for the New Jim Code*. Medford, MA: Polity.

Berinato, Scott. 2019. "Data Science and the Art of Persuasion." *Harvard Business Review* (January–February): 126–137. https://hbr.org/2019/01/data-science-and-the-art-of-persuasion.

Bica, Melissa, Julie L. Demuth, James E. Dykes, and Leysia Palen. 2019. "Communicating Hurricane Risks: Multi-Method Examination of Risk Imagery Diffusion." In *Proceedings of the 2019 CHI Conference on Human Factors in Computing Systems* (CHI '19), 1–13. Glasgow: Association for Computing Machinery. https://doi.org/10.1145/3290605.3300545.

Biles, John A. 2002. "Genjam: Evolutionary Computation Gets a Gig." In *Proceedings of the 2002 Conference for Information Technology Curriculum*. Rochester, NY: Society for Information Technology Education.

Bilis, Hélène. 2018. "Mapping Fiction: Social Networks and the Novel." Presentation at Shifting (the) Boundaries Conference, Wellesley College, April 6–7.

Bishop, Christopher M. 2006. *Pattern Recognition and Machine Learning*. Information Science and Statistics. New York: Springer.

Blei, David M., Andrew Y. Ng, and Michael I. Jordan. 2003. "Latent Dirichlet Allocation." *Journal of Machine Learning Research* 3 (January): 993–1022.

Blumenstock, Joshua, Gabriel Cadamuro, and Robert On. 2015. "Predicting Poverty and Wealth from Mobile Phone Metadata." *Science* 350 (6264): 1073–1076. https://doi.org/10.1126/science.aac4420.

Bødker, Susanne, Pelle Ehn, Joergen Knudsen, Morten Kyng, and Kim Madsen. 1988. "Computer Support for Cooperative Design." In *Proceedings of the 1988 ACM Conference on Computer-Supported Cooperative Work* (CSCW '88), 377–394. Portland, OR: Association for Computing Machinery.

Bolstad, William M., and James M. Curran. 2016. *Introduction to Bayesian Statistics*. Hoboken, NJ: Wiley.

Borgman, Christine L., Jillian C. Wallis, and Matthew S. Mayernik. 2012. "Who's Got the Data? Interdependencies in Science and Technology Collaborations." *Computer Supported Cooperative Work (CSCW)* 21 (6): 485–523. https://doi.org/10.1007/s10606-012-9169-z.

Borning, Alan, and Michael Muller. 2012. "Next Steps for Value Sensitive Design." In *Proceedings of the 2012 SIGCHI Conference on Human Factors in Computing Systems* (CHI '12), 1125–1134. Austin, TX: Association for Computing Machinery. https://doi.org/10.1145/2207676.2208560.

Bourgeois-Doyle, Dick. 2019. "Two-Eyed AI: A Reflection on Artificial Intelligence." Canadian Commission for UNESCO's IdeaLab. https://en.ccunesco.ca/-/media/Files/Unesco/Resources/2019/03/TwoEyedArtificial Intelligence.pdf.

Bowker, Geoffrey C., and Susan Leigh Star. 1998. "Building Information Infrastructures for Social Worlds— The Role of Classifications and Standards." In *Community Computing and Support Systems*, edited by Toru Ishida, 231–248. Berlin: Springer.

Bowker, Geoffrey C., and Susan Leigh Star. 2000. *Sorting Things Out: Classification and Its Consequences.* Cambridge, MA: MIT Press.

Brackeen, Brian. 2018. "Facial Recognition Software Is Not Ready for Use by Law Enforcement." *Tech Crunch*, June 25. https://techcrunch.com/2018/06/25/facial-recognition-software-is-not-ready-for-use-by-law -enforcement/.

Bradley, Adam, Cayley MacArthur, Mark Hancock, and Sheelagh Carpendale. 2015. "Gendered or Neutral? Considering the Language of HCI." In *Proceedings of the 41st Graphics Interface Conference*, 163–170. Halifax, Nova Scotia: Association for Computing Machinery.

Brichetto, Giampaolo, Margherita Monti Bragadin, Samuele Fiorini, Mario Alberto Battaglia, Giovanna Konrad, Michela Ponzio, Ludovico Pedullà, Alessandro Verri, Annalisa Barla, and Andrea Tacchino. 2020. "The Hidden Information in Patient-Reported Outcomes and Clinician-Assessed Outcomes: Multiple Sclerosis as a Proof of Concept of a Machine Learning Approach." *Neurological Sciences* 41 (2): 459–462. https://doi.org/10 .1007/s10072-019-04093-x.

Brooks, Michael, Kit Kuksenok, Megan K. Torkildson, Daniel Perry, John J. Robinson, Taylor J. Scott, Ona Anicello, Ariana Zukowski, Paul Harris, and Cecilia R. Aragon. 2013. "Statistical Affect Detection in Collaborative Chat." In *Proceedings of the 2013 Conference on Computer Supported Cooperative Work* (CSCW '13), 317–328. San Antonio, TX: Association for Computing Machinery. https://doi.org/10.1145/2441776.2441813.

Broussard, Meredith. 2018. *Artificial Unintelligence: How Computers Misunderstand the World.* Cambridge, MA: MIT Press.

Bruckman, Amy S., Casey Fiesler, Jeff Hancock, and Cosmin Munteanu. 2017. "CSCW Research Ethics Town Hall: Working towards Community Norms." In *Companion of the 2017 ACM Conference on Computer Supported Cooperative Work and Social Computing* (CSCW '17), 113–115. Portland, OR: Association for Computing Machinery. https://doi.org/10.1145/3022198.3022199.

Buolamwini, Joy. 2016. "How I'm Fighting Bias in Algorithms." TEDxBeacon Street. https://www.ted.com /talks/joy_buolamwini_how_i_m_fighting_bias_in_algorithms.

Buolamwini, Joy, and Timnit Gebru. 2018. "Gender Shades: Intersectional Accuracy Disparities in Commercial Gender Classification." *PMLR: Proceedings of the 1st Conference on Fairness, Accountability, and Transparency* 81: 77–91. http://proceedings.mlr.press/v81/buolamwini18a.html.

Burri, Regula Valérie, and Joseph Dumit. 2008. "Social Studies of Scientific Imaging and Visualization." In *The Handbook of Science and Technology Studies*, edited by Edward J. Hackett, Olga Amsterdamska, Michael Lynch, and Judy Wajcman, 297–317. Cambridge, MA: MIT Press.

Cabra, Mar. 2017. "How We Built the Data Team behind the Panama Papers." *Source*, November 29. https:// source.opennews.org/articles/how-we-built-data-team-behind-panama-papers/.

Carroll, Jennie. 2004. "Completing Design in Use: Closing the Appropriation Cycle," *Proceedings of 2004 European Conference on Information Systems* (ECIS 2004), 1–12. Turku, Finland: Association of Information Systems. https://aisel.aisnet.org/ecis2004/44/.

Centre for Social Justice and Community Action, Durham University, and National Co-ordinating Centre for Public Engagement. 2012. "Community-Based Participatory Research: A Guide to Ethical Principles and Practice." https://www.publicengagement.ac.uk/sites/default/files/publication/cbpr_ethics_guide_web_november _2012.pdf.

Chapelle, Olivier, Bernhard Schölkopf, and Alexander Zien. 2010. *Semi-Supervised Learning.* Cambridge, MA: MIT Press.

Charmaz, Kathy. 2006. *Constructing Grounded Theory: A Practical Guide through Qualitative Analysis.* London: SAGE Publications.

Chen, Nan-Chen, Margaret Drouhard, Rafal Kocielnik, Jina Suh, and Cecilia R. Aragon. 2018. "Using Machine Learning to Support Qualitative Coding in Social Science: Shifting the Focus to Ambiguity." *ACM Transactions on Interactive Intelligent Systems (TiiS)* 8 (2): 1–20.

Cheon, EunJeong, and Norman Makoto Su. 2018. "Futuristic Autobiographies: Weaving Participant Narratives to Elicit Values around Robots." In *Proceedings of the 2018 ACM/IEEE International Conference on*

Human-Robot Interaction (HRI '18), 388–397. Chicago: Association for Computing Machinery. https://doi.org /10.1145/3171221.3171244.

Chowdhury, Sudipta, Adindu Emelogu, Mohammad Marufuzzaman, Sarah G. Nurre, and Linkan Bian. 2017. "Drones for Disaster Response and Relief Operations: A Continuous Approximation Model." *International Journal of Production Economics* 188 (June): 167–184. https://doi.org/10.1016/j.ijpe.2017.03.024.

Cifor, Marika, Patricia Garcia, T. L. Cowan, Jasmine Rault, Tonia Sutherland, Anita Say Chan, Jennifer Rode, Anna Lauren Hoffman, Niloufar Salehi, and Lisa Nakamura. 2019. "Feminist Data Manifest-No." https://www .manifestno.com/.

Clark, Elizabeth, Anne Spencer Ross, Chenhao Tan, Yangfeng Ji, and Noah A. Smith. 2018. "Creative Writing with a Machine in the Loop: Case Studies on Slogans and Stories." In *Proceedings of the 23rd International Conference on Intelligent User Interfaces* (IUI '18), 329–340. Tokyo: Association for Computing Machinery. https://doi.org/10.1145/3172944.3172983.

Clarke, Victoria, Virginia Braun, and Nikki Hayfield. 2015. "Thematic Analysis." In *Qualitative Psychology: A Practical Guide to Research Methods*, edited by Jonathan Smith, 222–248. London: SAGE Publications.

Collins, Harry. 1985. *Changing Order: Replication and Induction in Scientific Practice*. Chicago: University of Chicago Press.

Collins, Patricia Hill. 2015. "Intersectionality's Definitional Dilemmas." *Annual Review of Sociology* 41: 1–20.

Cook, Karon F., Sally E. Jensen, Benjamin D. Schalet, Jennifer L. Beaumont, Dagmar Amtmann, Susan Cza- jkowski, Darren A. Dewalt, James F. Fries, Paul A. Pilkonis, Bryce B. Reeve et al. 2016. "PROMIS Measures of Pain, Fatigue, Negative Affect, Physical Function, and Social Function Demonstrated Clinical Validity across a Range of Chronic Conditions." *Journal of Clinical Epidemiology* 73 (May): 89–102. https://doi.org/10 .1016/j.jclinepi.2015.08.038.

Corbin, Juliet, and Anselm Strauss. 2014. *Basics of Qualitative Research: Techniques and Procedures for Developing Grounded Theory*. Los Angeles: SAGE Publications.

Corder, Gregory W., and Dale I. Foreman. 2014. *Nonparametric Statistics: A Step-by-Step Approach*. Hoboken, NJ: Wiley.

Correll, Michael. 2019. "Ethical Dimensions of Visualization Research." In *Proceedings of the 2019 CHI Con- ference on Human Factors in Computing Systems* (CHI '19), 1–13. Glasgow: Association for Computing Machinery. https://doi.org/10.1145/3290605.3300418.

Costanza-Chock, Sasha. 2020. *Design Justice: Community-Led Practices to Build the Worlds We Need*. Cam- bridge, MA: MIT Press.

Csikszentmihalyi, Mihaly. 1992. *Flow: The Psychology of Happiness*. New York: Random House.

Damasio, Antonio R. 2012. *Self Comes to Mind: Constructing the Conscious Brain*. New York: Vintage Books.

Daniels, Jessie, Karen Gregory, and Tressie McMillan Cottom, eds. 2016. *Digital Sociology in Everyday Life*. Bristol: Policy Press/Bristol University Press.

Davis, Murray S. 1971. "That's Interesting! Towards a Phenomenology of Sociology and a Sociology of Phe- nomenology." *Philosophy of the Social Sciences* 1 (2): 309–344. https://doi.org/10.1177/004839317100100211.

DeCamp, Abbie Levesque. 2020. "XM: A Schema for Encoding Queer Identities in Qualitative Research." *Computers and Composition* 55 (March): 102553. https://doi.org/10.1016/j.compcom.2020.102553.

DeChoudhury, Munmun, Michael Gamon, Scott Counts, and Eric Horvitz. 2013. "Predicting Depression via Social Media." In *Seventh International AAAI Conference on Weblogs and Social Media*. https://www.aaai.org /ocs/index.php/ICWSM/ICWSM13/paper/view/6124.

Degani, Asaf. 2004. *Taming Hal: Designing Interfaces Beyond 2001*. New York: Palgrave Macmillan.

Demner-Fushman, Dina, Wendy W. Chapman, and Clement J. McDonald. 2009. "What Can Natural Language Processing Do for Clinical Decision Support?" *Journal of Biomedical Informatics* 42 (5): 760–772. https://doi .org/10.1016/j.jbi.2009.08.007.

Dencik, Lina, Arne Hintz, Joanna Redden, and Emiliano Treré. 2019. "Exploring Data Justice: Conceptions, Applications, and Directions." *Information, Communication & Society* 22 (7): 873–881. https://doi.org/10.1080 /1369118X.2019.1606268.

Denham, Hannah. 2020. "IBM's Decision to Abandon Facial Recognition Technology Fueled by Years of Debate." *Washington Post*, June 11.

D'Ignazio, Catherine, and Lauren F. Klein. 2020. *Data Feminism*. Cambridge, MA: MIT Press.

Dilsizian, Steven E., and Eliot L. Siegel. 2013. "Artificial Intelligence in Medicine and Cardiac Imaging: Har- nessing Big Data and Advanced Computing to Provide Personalized Medical Diagnosis and Treatment." *Cur- rent Cardiology Reports* 16 (1): 441. https://doi.org/10.1007/s11886-013-0441-8.

Dix, Alan. 2007. "Designing for Appropriation." In *Proceedings of the 21st British HCI Group Annual Confer- ence on People and Computers: HCI . . . but Not as We Know It, Volume 2* (BCS-HCI '07), 27–30. University of Lancaster, UK: BCS Learning & Development.

Dossick, Carrie, Laura Osburn, and Gina Neff. 2019. "Innovation through Practice: The Messy Work of Making Technology Useful for Architecture, Engineering and Construction Teams." *Engineering, Construction and Architectural Management* (February). https://doi.org/10.1108/ECAM-12-2017-0272.

Dossick, Carrie Sturts, and Gina Neff. 2011. "Messy Talk and Clean Technology: Communication, Problem-Solving and Collaboration Using Building Information Modelling." *Engineering Project Organization Journal* 1 (2): 83–93. https://doi.org/10.1080/21573727.2011.569929.

Dossick, Carrie Sturts, Anne Anderson, Rahman Azari, Josh Iorio, Gina Neff, and John E. Taylor. 2015. "Messy Talk in Virtual Teams: Achieving Knowledge Synthesis through Shared Visualizations." *Journal of Management in Engineering* 31 (1): A4014003. https://doi.org/10.1061/(ASCE)ME.1943-5479.0000301.

Dossick, Carrie Sturts, Gina Neff, and Hoda Homayouni. 2009. "The Realities of Building Information Modeling for Collaboration in the AEC Industry." In *Building a Sustainable Future: Proceedings of the 2009 Construction Research Congress*, 396–405. Seattle, WA: American Society of Civil Engineers.

Dourish, Paul. 2001. "Process Descriptions as Organisational Accounting Devices: The Dual Use of Workflow Technologies." In *Proceedings of the 2001 International ACM SIGGROUP Conference on Supporting Group Work* (GROUP '01), 52–60. Boulder, CO: Association for Computing Machinery. https://doi.org/10.1145/500286.500297.

Drozdal, Jaimie, Justin Weisz, Dakuo Wang, Gaurav Dass, Bingsheng Yao, Changruo Zhao, Michael Muller, Lin Ju, and Hui Su. 2020. "Trust in AutoML: Exploring Information Needs for Establishing Trust in Automated Machine Learning Systems." In *Proceedings of the 25th International Conference on Intelligent User Interfaces* (IUI '20), 297–307. Cagliari, Italy: Association for Computing Machinery. https://doi.org/10.1145/3377325.3377501.

Drucker, Johanna. 2014. *Graphesis: Visual Forms of Knowledge Production*. Cambridge, MA: Harvard University Press.

Duarte, Marisa Elena, Morgan Vigil-Hayes, Sandra Littletree, and Miranda Belarde-Lewis. 2019. "'Of Course, Data Can Never Fully Represent Reality': Assessing the Relationship between Indigenous Data and IK, TEK, and TK." *Human Biology* 91 (3), 163–178. https://doi.org/10.13110/humanbiology.91.3.03 .

Durkheim, Émile. (1897) 1979. *Suicide: A Study in Sociology*. Edited by George Simpson. New York: Free Press.

Durkheim, Émile. (1895) 2014. *The Rules of Sociological Method: And Selected Texts on Sociology and Its Method*. Edited by Steven Lukes. Translated by W. D. Halls. New York: Free Press.

Eagle, Nathan, Alex (Sandy) Pentland, and David Lazer. 2009. "Inferring Friendship Network Structure by Using Mobile Phone Data." *Proceedings of the National Academy of Sciences* 106 (36): 15274–15278. https://doi.org/10.1073/pnas.0900282106.

Eubanks, Virginia. 2018. *Automating Inequality: How High-Tech Tools Profile, Police, and Punish the Poor*. New York: St. Martin's.

Faiola, Anthony, and Chris Newlon. 2011. "Advancing Critical Care in the ICU: A Human-Centered Biomedical Data Visualization Systems." In *Ergonomics and Health Aspects of Work with Computers*, edited by Michelle M. Robertson, 119–128. Berlin: Springer. http://dx.doi.org/10.1007/978-3-642-21716-6_13.

Feinberg, Melanie. 2007. "Hidden Bias to Responsible Bias: An Approach to Information Systems Based on Haraway's Situated Knowledges." *Information Research* 12 (4): 12–14.

Feinberg, Melanie. 2017. "A Design Perspective on Data." In *Proceedings of the 2017 CHI Conference on Human Factors in Computing Systems* (CHI '17), 2952–2963. Denver, CO: Association for Computing Machinery. https://doi.org/10.1145/3025453.3025837.

Few, Stephen. 2009. *Now You See It: Simple Visualization Techniques for Quantitative Analysis*. Oakland, CA: Analytics Press.

Fiesler, Casey. 2019. "Ethical Considerations for Research Involving (Speculative) Public Data." *Proceedings of the ACM on Human–Computer Interaction* 3 (GROUP): 1–13. https://doi.org/10.1145/3370271.

Fiesler, Casey, and Nicholas Proferes. 2018. "'Participant' Perceptions of Twitter Research Ethics." *Social Media + Society* 4 (1): 1–14. https://doi.org/10.1177/2056305118763366.

Fiesler, Casey, Cliff Lampe, and Amy S. Bruckman. 2016. "Reality and Perception of Copyright Terms of Service for Online Content Creation." In *Proceedings of the 19th ACM Conference on Computer-Supported Cooperative Work & Social Computing* (CSCW '16), 1450–1461. San Francisco: Association for Computing Machinery. https://doi.org/10.1145/2818048.2819931.

Finholt, Thomas A., and Gary M. Olson. 1997. "From Laboratories to Collaboratories: A New Organizational Form for Scientific Collaboration." *Psychological Science* 8 (1): 28–36. https://doi.org/10.1111/j.1467-9280.1997.tb00540.x.

Fiore-Gartland, Brittany, and Gina Neff. 2015. "Communication, Mediation, and the Expectations of Data: Data Valences across Health and Wellness Communities." *International Journal of Communication* 9: 1466–1484. http://ijoc.org/index.php/ijoc/article/view/2830.

Flanagan, Mary, and Helen Nissenbaum. 2014. *Values at Play in Digital Games*. Cambridge, MA: MIT Press.

Fleischmann, Kenneth R. 2013. "Information and Human Values." *Synthesis Lectures on Information Concepts, Retrieval, and Services* 5 (5): 1–99. https://doi.org/10.2200/S00545ED1V01Y201310ICR031.

Fleischmann, Kenneth R., and William A. Wallace. 2005. "A Covenant with Transparency: Opening the Black Box of Models." *Communications of the ACM* 48 (5): 93–97. https://doi.org/10.1145/1060710.1060715.

Floyd, Christiane, Heinz Züllighoven, Reinhard Budde, and Reinhard Keil-Slawik, eds. 1992. *Software Development as Reality Construction*. Berlin: Springer. https://doi.org/10.1007/978-3-642-76817-0_10.

Ford, Heather. 2014. "Big Data and Small: Collaborations between Ethnographers and Data Scientists." *Big Data & Society* 1 (2). https://doi.org/10.1177/2053951714544337.

Foster, Holly, and John Hagan. 2015. "Punishment Regimes and the Multilevel Effects of Parental Incarceration: Intergenerational, Intersectional, and Interinstitutional Models of Social Inequality and Systemic Exclusion." *Annual Review of Sociology* 41: 135–158.

franzke, aline shakti. 2020. "Feminist Research Ethics, IRE 3.0 Companion 6.3." Association of Internet Researchers, 64–75. https://aoir.org/reports/ethics3.pdf.

franzke, aline shakti, Anja Bechmann, Michael Zimmer, Charles Ess, and the Association of Internet Researchers. 2020. "Internet Research: Ethical Guidelines 3.0." https://aoir.org/reports/ethics3.pdf.

Friedler, Sorelle A., Carlos Scheidegger, and Suresh Venkatasubramanian. 2016. "On the (Im)Possibility of Fairness." *ArXiv:1609.07236 [Cs.CY]*, September. http://arxiv.org/abs/1609.07236.

Friedman, Batya, and David G. Hendry. 2019. *Value Sensitive Design: Shaping Technology with Moral Imagination*. Cambridge, MA: MIT Press.

Friedman, Batya, Peter H. Kahn, and Alan Borning. 2008. "Value Sensitive Design and Information Systems." In *The Handbook of Information and Computer Ethics*, edited by Kenneth Einar Himma and Herman T. Tavani, 69–101. Hoboken, NJ: Wiley.

Geertz, Clifford. 1973. *The Interpretation of Cultures: Selected Essays*. New York: Basic Books.

Girasa, Rosario. 2020. "Bias, Jobs, and Fake News." In *Artificial Intelligence as a Disruptive Technology*, 187–215. Cham, Switzerland: Springer International. https://doi.org/10.1007/978-3-030-35975-1_6.

Glaser, Barney G., and Anselm L. Strauss. 1967. *The Discovery of Grounded Theory: Strategies for Qualitative Research*. Hawthorne, NY: Aldine de Gruytner.

Gluesing, Julia, Kenneth Riopelle, and James Danowski. 2014. "Mixing Ethnography and Information Technology Data Mining to Visualize Innovation Networks in Global Networked Organizations." *Mixed Methods Social Networks Research: Design and Applications* 36: 203.

Goel, Ashok K., and Michael E. Helms. 2014. "Theories, Models, Programs, and Tools of Design: Views from Artificial Intelligence, Cognitive Science, and Human-Centered Computing." In *An Anthology of Theories and Models of Design*, edited by Amaresh Chakrabarti and Lucienne T. M. Blessing, 417–432. London: Springer.

González, Felipe, Yihan Yu, Andrea Figueroa, Claudia López, and Cecilia Aragon. 2019. "Global Reactions to the Cambridge Analytica Scandal: A Cross-Language Social Media Study." In *Companion Proceedings of the 2019 World Wide Web Conference* (WWW '19), 799–806. San Francisco: Association for Computing Machinery.

Gray, Mary L., and Siddharth Suri. 2019. *Ghost Work: How to Stop Silicon Valley from Building a New Global Underclass*. Boston: Houghton Mifflin Harcourt.

Gray, Wayne D., Bonnie E. John, and Michael E. Atwood. 1993. "Project Ernestine: Validating a GOMS Analysis for Predicting and Explaining." *Human–Computer Interaction* 8: 237–309. https://doi.org/10.1207/s15327051hci0803_3.

Gray, Wayne D., Bonnie E. John, Rory Stuart, Deborah Lawrence, and Michael E. Atwood. 1995. "GOMS Meets the Phone Company: Analytic Modeling Applied to Real-World Problems." In *Readings in Human–Computer Interaction*, edited by Ronald M. Baecker, Jonathan Grudin, and Saul Greenberg, 634–639. San Francisco, CA: Morgan Kaufmann Publishers. https://doi.org/10.1016/B978-0-08-051574-8.50065-0.

Green, Tera Marie, Richard Arias-Hernandez, and Brian Fisher. 2014. "Individual Differences and Translational Science in the Design of Human-Centered Visualizations." In *Handbook of Human Centric Visualization*, edited by Weidong Huang, 93–113. New York: Springer. https://doi.org/10.1007/978-1-4614-7485-2_4.

Guha, Shion, Michael Muller, N. Sadat Shami, Mikhil Masli, and Werner Geyer. 2016. "Using Organizational Social Networks to Predict Employee Engagement." In *Tenth International AAAI Conference on Web and Social Media* (ICWSM 2016), 131–136. Cologne: Association for the Advancement of Artificial Intelligence. https://www.aaai.org/ocs/index.php/ICWSM/ICWSM16/paper/view/13157.

Guo, Philip Jia. 2012. "Software Tools to Facilitate Research Programming." PhD diss., Stanford University.

Hallinan, Blake, and Ted Striphas. 2016. "Recommended for You: The Netflix Prize and the Production of Algorithmic Culture." *New Media & Society* 18 (1): 117–137. https://doi.org/10.1177/1461444814538646.

Handelman, G. S., H. K. Kok, R. V. Chandra, A. H. Razavi, S. Huang, M. Brooks, M. J. Lee. and H. Asadi. 2019. Peering into the Black Box of Artificial Intelligence: Evaluation Metrics of Machine Learning Methods. *American Journal of Roentgenology* 212 (1): 38–43.

Haraway, Donna. 1988. "Situated Knowledges: The Science Question in Feminism and the Privilege of Partial Perspective." *Feminist Studies* 14 (3): 575–599. https://doi.org/10.2307/3178066.

Harding, Sandra G., ed. 2004. *The Feminist Standpoint Theory Reader: Intellectual and Political Controversies.* New York: Routledge.

Hayes, Bob. 2018. "Top 10 Challenges to Practicing Data Science at Work." *BusinessOverBroadway*, March 18.

Hayes, Gillian R. 2011. "The Relationship of Action Research to Human–Computer Interaction." *ACM Transactions on Computer–Human Interaction* 18 (3): 1–20. https://doi.org/10.1145/1993060.1993065.

Hemsley, Jeff, and Jaime Snyder. 2018. "Dimensions of Visual Misinformation in the Emerging Media Landscape." In *Misinformation and Mass Audiences*, edited by Brian Southwell, Emily A. Thorson and Laura Sheble, 91–108. Austin, TX: University of Texas Press.

Hill, Benjamin Mako, Dharma Dailey, Richard T. Guy, Ben Lewis, Mika Matsuzaki, and Jonathan T. Morgan. 2017. "Democratizing Data Science: The Community Data Science Workshops and Classes." In *Big Data Factories: Collaborative Approaches*, edited by Sorin Adam Matei, Nicolas Jullien, and Sean P. Goggins, 115–135. Computational Social Sciences. Cham, Switzerland: Springer International. https://doi.org/10.1007/978-3-319 -59186-5_9.

Hirsch, Tad. 2011. "More than Friends: Social and Mobile Media for Activist Organizations." In *From Social Butterfly to Engaged Citizen: Urban Informatics, Social Media, Ubiquitous Computing, and Mobile Technology to Support Citizen Engagement*, edited by Marcus Foth, Laura Forlano, Christine Satchell, and Martin Gibbs, 135–150. Cambridge, MA: MIT Press.

Houde, Stephanie, Vera Liao, Jacquelyn Martino, Michael Muller, David Piorkowski, John Richards, Justin Weisz, and Yunfeng Zhang. 2020. "Business (Mis)Use Cases of Generative AI." In *Proceedings of IUI 2020 Workshop on Human-AI Co-Creation with Generative Models*. New York: Association for Computing Machinery.

Howard, Philip. 2002. "Network Ethnography and the Hypermedia Organization: New Media, New Organizations, New Methods." *New Media & Society* 4: 550–574. https://doi.org/10.1177/146144402321466813.

Howard, Philip. 2020. *Lie Machines: How to Save Democracy from Troll Armies, Deceitful Robots, Junk News Operations, and Political Operatives*. New Haven, CT: Yale University Press.

Huang, Cheng-Zhi Anna, Curtis Hawthorne, Adam Roberts, Monica Dinculescu, James Wexler, Leon Hong, and Jacob Howcroft. 2019. "The Bach Doodle: Approachable Music Composition with Machine Learning at Scale." *ArXiv:1907.06637 [Cs.SD]*, July. http://arxiv.org/abs/1907.06637.

Huang, Jingwei. 2018. "From Big Data to Knowledge: Issues of Provenance, Trust, and Scientific Computing Integrity." In *2018 IEEE International Conference on Big Data (Big Data)*, 2197–2205. IEEE.

Iliadis, Andrew, and Federica Russo. 2016. "Critical Data Studies: An Introduction." *Big Data & Society* 3 (2), 1–7. https://doi.org/10.1177/2053951716674238.

Institute for the Future and Omidyar Network Tech and Society Solutions Lab. 2018. "Ethical OS." https:// ethicalos.org/.

Institute for Health Metrics and Evaluation. 2020. "COVID-19 Projections." IHME. https://covid19.healthdata .org/.

Kandel, Sean, Andreas Paepcke, Joseph M. Hellerstein, and Jeffrey Heer. 2012. "Enterprise Data Analysis and Visualization: An Interview Study." *IEEE Transactions on Visualization and Computer Graphics* 18 (12): 2917–2926.

Kanter, James Max, and Kalyan Veeramachaneni. 2015. "Deep Feature Synthesis: Towards Automating Data Science Endeavors." In *2015 IEEE International Conference on Data Science and Advanced Analytics (DSAA)*, 1–10. https://doi.org/10.1109/DSAA.2015.7344858.

Kassambara, Alboukadel. 2017. *Practical Guide to Cluster Analysis in R: Unsupervised Machine Learning*. CreateSpace Independent Publishing.

Kay, Matthew, Steve Haroz, Shion Guha, and Pierre Dragicevic. 2016. "Special Interest Group on Transparent Statistics in HCI." In *Proceedings of the 2016 CHI Conference Extended Abstracts on Human Factors in Computing Systems* (CHI EA '16), 1081–1084. San Jose, CA: Association for Computing Machinery. https://doi.org /10.1145/2851581.2886442.

Kay, Matthew, Steve Haroz, Shion Guha, Pierre Dragicevic, and Chat Wacharamanotham. 2017. "Moving Transparent Statistics Forward at CHI." In *Proceedings of the 2017 CHI Conference Extended Abstracts on Human Factors in Computing Systems* (CHI EA '17), 534–541. Denver, CO: Association for Computing Machinery. https://doi.org/10.1145/3027063.3027084.

Kery, Mary Beth. 2018. "Towards Scaffolding Complex Exploratory Data Science Programming Practices." In *2018 IEEE Symposium on Visual Languages and Human-Centric Computing (VL/HCC)*, 273–274. IEEE. https://doi.org/10.1109/VLHCC.2018.8506555.

Kery, Mary Beth, Marissa Radensky, Mahima Arya, Bonnie E John, and Brad A Myers. 2018. "The Story in the Notebook: Exploratory Data Science Using a Literate Programming Tool." In *Proceedings of the 2018 CHI Conference on Human Factors in Computing Systems* (CHI '18), 174. Montreal: Association for Computing Machinery. https://doi.org/10.1145/3173574.3173748.

Kerzner, Ethan, Sarah Goodwin, Jason Dykes, Sara Jones, and Miriah Meyer. 2018. "A Framework for Creative Visualization–Opportunities Workshops." *IEEE Transactions on Visualization and Computer Graphics* 25: 748–758.

Kim, Miryung, Thomas Zimmermann, Robert DeLine, and Andrew Begel. 2016. "The Emerging Role of Data Scientists on Software Development Teams." In *2016 IEEE/ACM 38th International Conference on Software Engineering* (ICSE '16), 96–107. Austin, TX: Association for Computing Machinery. https://doi.org/10.1145/2884781.2884783.

King, Gary, Jennifer Pan, and Margaret E. Roberts. 2013. "How Censorship in China Allows Government Criticism but Silences Collective Expression." *American Political Science Review* 107 (02): 326–343. https://doi.org/10.1017/S0003055413000014.

King, Gary, Jennifer Pan, and Margaret E. Roberts. 2017. "How the Chinese Government Fabricates Social Media Posts for Strategic Distraction, Not Engaged Argument." *American Political Science Review* 111 (3): 484–501. https://doi.org/10.1017/S0003055417000144.

Kitchin, Rob, and Tracey Lauriault. 2018. "Towards Critical Data Studies: Charting and Unpacking Data Assemblages and Their Work." In *Think Big Data in Geography: New Regimes, New Research*, edited by Jim Thatcher, Josef Eckert and Andrew Shears, 3–20. Lincoln: University of Nebraska Press.

Knaflic, Cole Nussbaumer. 2015. *Storytelling with Data: A Data Visualization Guide for Business Professionals*. Hoboken, NJ: Wiley.

Knuth, D. E. 1984. "Literate Programming." *The Computer Journal* 27 (2): 97–111. https://doi.org/10.1093/comjnl/27.2.97.

Koepfler, Jes A., and Kenneth R. Fleischmann. 2012. "Studying the Values of Hard-to-Reach Populations: Content Analysis of Tweets by the 21st Century Homeless." In *Proceedings of the 2012 iConference*, (iConference '12): 48–55. Toronto: Association for Computing Machinery.

Kogan, Marina, Jennings Anderson, Leysia Palen, Kenneth M. Anderson, and Robert Soden. 2016. "Finding the Way to OSM Mapping Practices: Bounding Large Crisis Datasets for Qualitative Investigation." In *Proceedings of the 2016 CHI Conference on Human Factors in Computing Systems* (CHI '16), 2783–2795. San Jose, CA: Association for Computing Machinery.

Kogan, Marina, Leysia Palen, and Kenneth M. Anderson. 2015. "Think Local, Retweet Global: Retweeting by the Geographically-Vulnerable during Hurricane Sandy." In *Proceedings of the 18th ACM Conference on Computer Supported Cooperative Work & Social Computing* (CSCW '15), 981–993. Vancouver: Association for Computing Machinery. https://doi.org/10.1145/2675133.2675218.

Krischkowsky, Alina, Manfred Tscheligi, Katja Neureiter, Michael Muller, Anna Maria Polli, and Nervo Verdezoto. 2015. "Experiences of Technology Appropriation: Unanticipated Users, Usage, Circumstances, and Design." *Proceedings of the 14th European Conference on Computer Supported Cooperative Work*. https://uni-salzburg.elsevierpure.com/en/publications/experiences-of-technology-appropriation-unanticipated-users-usage.

Krotov, Vlad, and Leiser Silva. 2018. "Legality and Ethics of Web Scraping." *AMCIS 2018 Proceedings* (August). https://aisel.aisnet.org/amcis2018/DataScience/Presentations/17.

Kugler, Kari C., John J. Dziak, and Jessica Trail. 2018. "Coding and Interpretation of Effects in Analysis of Data from a Factorial Experiment." In *Optimization of Behavioral, Biobehavioral, and Biomedical Interventions: Advanced Topics*, edited by Linda M. Collins and Kari C. Kugler, 175–205. Statistics for Social and Behavioral Sciences. Cham, Switzerland: Springer International. https://doi.org/10.1007/978-3-319-91776-4_6.

Kuksenok, Kit. 2016. "Influence Apart from Adoption: How Interaction between Programming and Scientific Practices Shapes Modes of Inquiry in Four Oceanography Teams." PhD diss., University of Washington. https://digital.lib.washington.edu/researchworks/handle/1773/36546.

Kuksenok, Kit, Cecilia Aragon, James Fogarty, Charlotte P. Lee, and Gina Neff. 2017. "Deliberate Individual Change Framework for Understanding Programming Practices in Four Oceanography Groups." In *Proceedings of the 2017 ACM Conference on Computer-Supported Cooperative Work and Social Computing*, 663–691. (CSCW '17). Portland, OR: Association for Computing Machinery. https://doi.org/10.1007/s10606-017-9285-x.

Kyng, M., and L. Mathiassen, eds. 1995. *Computers in Context: Joining Forces in Design. The Third Decennial Conference*. Aarhus, Denmark: Department of Computer Science, Aarhus University.

Lasecki, Walter S., Juho Kim, Nick Rafter, Onkur Sen, Jeffrey P. Bigham, and Michael S. Bernstein. 2015. "Apparition: Crowdsourced User Interfaces That Come to Life as You Sketch Them." In *Proceedings of the*

33rd Annual ACM Conference on Human Factors in Computing Systems ('CHI 15), 1925–1934. Seoul: Association for Computing Machinery.

Lazer, David, Alex Pentland, Lada Adamic, Sinan Aral, Albert-László Barabási, Devon Brewer, Nicholas Christakis, Noshir Contractor, James Fowler, Myron Gutmann et al. 2009. "Computational Social Science." *Science* 323: 721–723.

Le Dantec, Christopher A., and Carl DiSalvo. 2013. "Infrastructuring and the Formation of Publics in Participatory Design." *Social Studies of Science* 43 (2): 241–264.

Le Dantec, Christopher A., Erika Shehan Poole, and Susan P Wyche. 2009. "Values as Lived Experience: Evolving Value Sensitive Design in Support of Value Discovery." In *Proceedings of the SIGCHI Conference on Human Factors in Computing Systems* (CHI '09), 1141–1150. Boston: Association for Computing Machinery.

Leslie, David. 2019. *Understanding Artificial Intelligence Ethics and Safety: A Guide for the Responsible Design and Implementation of AI Systems in the Public Sector.* London: The Alan Turing Institute. https://doi.org/10.5281/zenodo.3240529.

Lewis, Clayton, Peter G. Polson, Cathleen Wharton, and John Rieman. 1990. "Testing a Walkthrough Methodology for Theory-Based Design of Walk-up-and-Use Interfaces." In *Proceedings of the SIGCHI Conference on Human Factors in Computing Systems* (CHI '90), 235–242. Seattle: Association for Computing Machinery.

Li, Xuyang, Antara Bahursettiwar, and Marina Kogan. 2021. "Hello? Is There Anybody in There? Analysis of Factors Promoting Response from Authoritative Sources in Crisis." *Proceedings of the 2019 CSCW Conference on Computer Supported Cooperative Work and Social Computing 2021* 5 (1). https://dl.acm.org/doi/10.1145/3449209?sid=SCITRUS.

Liang, Jason, Elliot Meyerson, Babak Hodjat, Dan Fink, Karl Mutch, and Risto Miikkulainen. 2019. "Evolutionary Neural AutoML for Deep Learning." In *Proceedings of the Genetic and Evolutionary Computation Conference* (GECCO '19), 401–409. Prague, Czech Republic: Association for Computing Machinery. https://doi.org/10.1145/3321707.3321721.

Light, Ben, Jean Burgess, and Stefanie Duguay. 2018. "The Walkthrough Method: An Approach to the Study of Apps." *New Media & Society* 20 (3): 881–900. https://doi.org/10.1177/1461444816675438.

Lindman, Harold R. 2012. *Analysis of Variance in Experimental Design.* New York: Springer Science & Business Media.

Lowry, Richard. 2014. *Concepts and Applications of Inferential Statistics.* Accessed June 2, 2020. http://vassarstats.net/textbook/.

Lynch, Michael. 1988. "The Externalized Retina: Selection and Mathematization in the Visual Documentation of Objects in the Life Sciences." *Human Studies* 11: 201–234.

Luczak-Roesch, Markus, Ramine Tinati, Kieron O'Hara, and Nigel Shadbolt. 2015. "Socio-Technical Computation." In *Proceedings of the 18th ACM Conference Companion on Computer Supported Cooperative Work & Social Computing* (CSCW '15), 139–142. Vancouver: Association for Computing Machinery. https://doi.org/10.1145/2685553.2698991.

MacKenzie, Donald A., and Judy Wajcman. 1999. *The Social Shaping of Technology.* 2nd ed. Buckingham, UK: Open University Press.

Mackinlay, Jock. 1986. "Automating the Design of Graphical Presentations of Relational Information." *ACM Transactions on Graphics* 5 (2): 110–141. https://doi.org/10.1145/22949.22950.

Maglogiannis, Ilias G. 2007. *Emerging Artificial Intelligence Applications in Computer Engineering: Real Word AI Systems with Applications in EHealth, HCI, Information Retrieval, and Pervasive Technologies.* Amsterdam: IOS Press.

Magnus, Brooke E., Yang Liu, Jason He, Hally Quinn, David Thissen, Heather E. Gross, Darren A. DeWalt, and Bryce B. Reeve. 2016. "Mode Effects between Computer Self-Administration and Telephone Interviewer-Administration of the PROMIS® Pediatric Measures, Self- and Proxy Report." *Quality of Life Research* 25 (7): 1655–1665. https://doi.org/10.1007/s11136-015-1221-2.

Maitland, Sarah. 2017. *What Is Cultural Translation?* Bloomsbury Advances in Translation Series. London: Bloomsbury Academic.

Maitra, Suvradip. 2020. "Artificial Intelligence and Indigenous Perspectives: Protecting and Empowering Intelligent Human Beings." In *Proceedings of the AAAI/ACM Conference on AI, Ethics, and Society* (AIES '20), 320–326. New York: Association for Computing Machinery. https://doi.org/10.1145/3375627.3375845.

Mao, Yaoli, Dakuo Wang, Michael Muller, Kush R. Varshney, Ioana Baldini, Casey Dugan, and Aleksandra Mojsilović. 2019. "How Data Scientists Work Together with Domain Experts in Scientific Collaborations: To Find the Right Answer or to Ask the Right Question?" *Proceedings of the ACM on Human–Computer Interaction* 3 (GROUP): 1–23. https://doi.org/10.1145/3361118.

Markham, Annette N. 2006. "Method as Ethic, Ethic as Method." *Journal of Information Ethics* 15 (2): 37–55.

Markham, Annette N., and Nancy K. Baym, eds. 2009. *Internet Inquiry: Conversations about Method*. Los Angeles: SAGE Publications.

Matsudaira, Kate. 2015. "The Science of Managing Data Science." *Acmqueue* 13 (4). https://queue.acm.org /detail.cfm?id=2767971.

Mayernik, Matthew S., Tim DiLauro, Ruth Duerr, Elliot Metsger, Anne E. Thessen, and G. Sayeed Choud-hury. 2013. "Data Conservancy Provenance, Context, and Lineage Services: Key Components for Data Preser-vation and Curation." *Data Science Journal* 12: 158–171.

McKinney, Wes. 2018. *Python for Data Analysis: Data Wrangling with Pandas, NumPy, and IPython*. Sebas-topol, CA: O'Reilly.

Mentis, Helena M., Ahmed Rahim, and Pierre Theodore. 2016. "Crafting the Image in Surgical Telemedicine." In *Proceedings of the 19th ACM Conference on Computer-Supported Cooperative Work & Social Computing* (CSCW '16), 744–755. San Francisco: Association for Computing Machinery. https://doi.org/10.1145/2818048 .2819978.

Miller, Christopher A. 2014. "Delegation and Intent Expression for Human-Automation Interaction: Thoughts for Single Pilot Operations." In *Proceedings of the International Conference on Human–Computer Interaction in Aerospace*, (HCI-Aero '14), 1–10. Santa Clara, CA: Association for Computing Machinery. https://doi.org /10.1145/2669592.2669649.

Miller, Daniel, Elisabetta Costa, Neil Haynes, Tom McDonald, Razvan Nicolescu, Jolynna Sinanan, Juliano Spyer, and Shriram Venkatraman. 2016. *How the World Changed Social Media*. London: UCL Press.

Minervino, Ricardo A, Alejandra Martín, L Micaela Tavernini, and Máximo Trench. 2018. "The Understand-ing of Visual Metaphors by the Congenitally Blind." *Frontiers in Psychology* 9: 1242.

Mitchell, Shira, Eric Potash, Solon Barocas, Alexander D'Amour, and Kristian Lum. 2019. "Prediction-Based Decisions and Fairness: A Catalogue of Choices, Assumptions, and Definitions." *ArXiv:1811.07867 [Stat.AP]*, July. http://arxiv.org/abs/1811.07867.

Mitra, Tanushree, Michael Muller, N. Sadat Shami, Abbas Golestani, and Mikhil Masli. 2017. "Spread of Employee Engagement in a Large Organizational Network: A Longitudinal Analysis." In *Proceedings of the ACM on Human–Computer Interaction* 1 (CSCW): 1–20.

Møller, Naja L. Holten, Kathleen H. Pine, Claus Bossen, Trine Nielsen, and Gina Neff. 2020. "Who Does the Work of Data?" *ACM Interactions* 27 (3) 52–55. https://dl.acm.org/doi/10.1145/3386389.

Molnar, Christoph. 2020. *Interpretable Machine Learning*. Morrisville, NC: Lulu.

Monge, Peter R. and Noshir S. Contractor. 2003. *Theories of Communication Networks*. New York: Oxford University Press.

Muller, Michael J. 1997. "Ethnocritical Heuristics for Reflecting on Work with Users and Other Interested Par-ties." In *Computers and Design in Context*, edited by Morten Kyng, 349–380. Cambridge, MA: MIT Press.

Muller, Michael J. 1999. "Invisible Work of Telephone Operators: An Ethnocritical Analysis." *Computer Sup-ported Cooperative Work (CSCW)* 8 (1–2): 31–61.

Muller, Michael J. 2001. "Layered Participatory Analysis: New Developments in the CARD Technique." In *Proceedings of the SIGCHI Conference on Human Factors in Computing Systems* (CHI '01), 90–97. Seattle: Association for Computing Machinery.

Muller, Michael J. 2011. "Feminism Asks the 'Who' Questions in HCI." *Interacting with Computers* 23 (5): 447–449.

Muller, Michael J. 2014. "Curiosity, Creativity, and Surprise as Analytic Tools: Grounded Theory Method." In *Ways of Knowing in HCI*, edited by Judith Olson and Wendy Kellogg, 25–48. New York: Springer.

Muller, Michael J., and Allison Druin. 2012. "Participatory Design: The Third Space in HCI." In *The Human– Computer Interaction Handbook*, edited by Julia A. Jacko, 3rd ed. New York: Taylor and Francis.

Muller, Michael J., and Kenneth Carey. 2002. "Design as a Minority Discipline in a Software Company: Toward Requirements for a Community of Practice." In *Proceedings of the SIGCHI Conference on Human Factors in Computing Systems* (CHI '02), 383–390. Minneapolis: Association for Computing Machinery.

Muller, Michael, Rebecca Carr, Catherine Ashworth, Barbara Diekmann, Cathleen Wharton, Cherie Eicks-taedt, and Joan Clonts. 1995. "Telephone Operators as Knowledge Workers: Consultants Who Meet Customer Needs. In Proceedings of the CHI 1995 Conference on Human Factors in Computing Systems (CHI '95), 130–137. Denver: Association for Computing Machinery. https://doi.org/10.1145/223904.223921.

Muller, Michael, Christine T. Wolf, Josh Andres, Michael Desmond, Narendra Nath Joshi, Zahra Ashktorab, Aabhas Sharma, Kristina Brimijoin, Qian Pan, Evelyn Duesterwald, and Casey Dugan. 2021. "Designing Ground Truth and the Social Life of Labels." In *Proceedings of the SIGCHI Conference on Human Factors in Computing Systems* (CHI '21), 1–16. New York: Association for Computing Machinery. https://doi.org/10.1145 /3411764.3445402.

Muller, Michael, and Q. Vera Liao. 2017. "Exploring AI Ethics and Values through Participatory Design Fictions." *Human Computer Interaction Consortium.* http://www.qveraliao.com/hcic2017.pdf.

Muller, Michael, and Sacha Chua. 2012. "Brainstorming for Japan: Rapid Distributed Global Collaboration for Disaster Response." In *Proceedings of the SIGCHI Conference on Human Factors in Computing Systems* (CHI '12), 2727–2730. Austin, TX: Association for Computing Machinery. https://doi.org/10.1145/2207676.2208668.

Muller, Michael, and Thomas Erickson. 2018. "In the Data Kitchen: A Review (a Design Fiction on Data Science)." In *Extended Abstracts of the 2018 CHI Conference on Human Factors in Computing Systems* (CHI EA '18), 1–10. Montreal: Association for Computing Machinery. https://doi.org/10.1145/3170427.3188407.

Muller, Michael, Ingrid Lange, Dakuo Wang, David Piorkowski, Jason Tsay, Q. Vera Liao, Casey Dugan, and Thomas Erickson. 2019. "How Data Science Workers Work with Data: Discovery, Capture, Curation, Design, Creation." In *Proceedings of the 2019 CHI Conference on Human Factors in Computing Systems* (CHI '19), 1–15. Glasgow: Association for Computing Machinery. https://doi.org/10.1145/3290605.3300356.

Muller, Michael, Melanie Feinberg, Timothy George, Steven J. Jackson, Bonnie E. John, Mary Beth Kery, and Samir Passi. 2019. "Human-Centered Study of Data Science Work Practices." In *Extended Abstracts of the 2019 CHI Conference on Human Factors in Computing Systems* (CHI EA '19), 1–8. Glasgow: Association for Computing Machinery. https://doi.org/10.1145/3290607.3299018.

Muller, Michael, N. Sadat Shami, Shion Guha, Mikhil Masli, Werner Geyer, and Alan Wild. 2016. "Influences of Peers, Friends, and Managers on Employee Engagement." In *Proceedings of the 19th International Conference on Supporting Group Work* (GROUP '16), 131–136. Sanibel Island, FL: Association for Computing Machinery.

Muller, Michael, Shion Guha, Eric P.S. Baumer, David Mimno, and N. Sadat Shami. 2016. "Machine Learning and Grounded Theory Method: Convergence, Divergence, and Combination." In *Proceedings of the 19th International Conference on Supporting Group Work* (GROUP '16), 3–8. Sanibel Island, FL: Association for Computing Machinery.

Murnane, Elizabeth L., and Scott Counts. 2014. "Unraveling Abstinence and Relapse: Smoking Cessation Reflected in Social Media." In *Proceedings of the SIGCHI Conference on Human Factors in Computing Systems* (CHI '14), 1345–1354. Toronto: Association for Computing Machinery. https://doi.org/10.1145/2556288.2557145.

Murthy, Dhiraj. 2011. "Emergent Digital Ethnographic Methods for Social Research." In *Handbook of Emergent Technologies in Social Research*, edited by Sharlene Nagy Hesse-Biber, 158–179. Oxford: Oxford University Press.

Nagy, Peter, and Gina Neff. 2015. "Imagined Affordances: Reconstructing a Keyword for Communication Theory." *Social Media + Society* 1 (2): 1–9. https://doi.org/10.1177/2056305115603385.

Narayanan, Arvind. 2018. "21 Fairness Definitions and Their Politics." Tutorial at ACM FAccT Conference. New York: ACM. https://facctconference.org/2018/livestream_vh220.html

National Commission for the Protection of Human Subjects of Biomedical and Behavioral Research. 1979. "The Belmont Report: Ethical Principles and Guidelines for the Protection of Human Subjects of Research." Washington, D.C.: Department of Health, Education, and Welfare. https://www.hhs.gov/ohrp/sites/default/files/the-belmont-report-508c_FINAL.pdf.

National Co-ordinating Centre for Public Engagement and University of Bristol. 2017. "Working with Local Communities." https://www.publicengagement.ac.uk/sites/default/files/publication/working_with_local_communities.pdf.

Neff, Gina, and Dawn Nafus. 2016. *Self-Tracking.* Cambridge, MA: MIT Press.

Neff, Gina, Anissa Tanweer, Brittany Fiore-Gartland, and Laura Osburn. 2017. "Critique and Contribute: A Practice-Based Framework for Improving Critical Data Studies and Data Science." *Big Data* 5 (2): 85–97. https://doi.org/10.1089/big.2016.0050.

Newman, Mark. 2018. *Networks.* Oxford University Press.

Nguyen, Thien Hoang, Muqing Cao, Thien-Minh Nguyen, and Lihua Xie. 2018. "Post-Mission Autonomous Return and Precision Landing of UAV." *2018 15th International Conference on Control, Automation, Robotics and Vision (ICARCV)*, 1747–1752. Singapore: IEEE. https://doi.org/10.1109/ICARCV.2018.8581117.

Nielsen, Jakob. 1992. "Finding Usability Problems through Heuristic Evaluation." In *Proceedings of the SIGCHI Conference on Human Factors in Computing Systems* (CHI' 92), 373–380. Monterey, CA: Association for Computing Machinery.

Nissenbaum, Helen. 2001. "How Computer Systems Embody Values." *Computer* 34 (3): 118–120.

Noble, Safiya Umoja. 2018. *Algorithms of Oppression: How Search Engines Reinforce Racism.* New York: NYU Press.

Nobre, Carolina, Marc Streit, and Alexander Lex. 2018. "Juniper: A Tree+ Table Approach to Multivariate Graph Visualization." *IEEE Transactions on Visualization and Computer Graphics* 25 (1): 544–554.

Nobre, Carolina, Nils Gehlenborg, Hilary Coon, and Alexander Lex. 2019. "Lineage: Visualizing Multivariate Clinical Data in Genealogy Graphs." *IEEE Transactions on Visualization and Computer Graphics* 25 (3): 1543–1558.

Nunavut Department of Education. 2007. "Inuit Qaujimajatuqangit: Educational Framework for Nunavut Curriculum." Government of Nunavut. https://www.gov.nu.ca/sites/default/files/files/Inuit%20Qaujimajatuqangit%20ENG.pdf.

Nuzzo, Regina. 2014. "Scientific Method: Statistical Errors." *Nature News* 506 (7487): 150. https://doi.org/10.1038/506150a.

Oh, Changhoon, Jungwoo Song, Jinhan Choi, Seonghyeon Kim, Sungwoo Lee, and Bongwon Suh. 2018. "I Lead, You Help but Only with Enough Details: Understanding User Experience of Co-Creation with Artificial Intelligence." In *Proceedings of the 2018 CHI Conference on Human Factors in Computing Systems* (CHI '18), 1–13. Montreal: Association for Computing Machinery. https://doi.org/10.1145/3173574.3174223.

Ohana, Hana. n.d. "The Science of Collaboratories." Accessed April 27, 2020. http://hana.ics.uci.edu/index.html%3Fq=content%252Fscience-collaboratories.html.

Olson, Gary M., Ann Zimmerman, and Nathan Bos, eds. 2008. *Scientific Collaboration on the Internet.* Acting with Technology. Cambridge, MA: MIT Press.

Olson, Randal S., and Jason H. Moore. 2018. "Identifying and Harnessing the Building Blocks of Machine Learning Pipelines for Sensible Initialization of a Data Science Automation Tool." In *Genetic Programming Theory and Practice XIV,* edited by Rick Riolo, Bill Worzel, Brian Goldman, and Bill Tozier, 211–23. Genetic and Evolutionary Computation. Cham, Switzerland: Springer International. https://doi.org/10.1007/978-3-319-97088-2_14.

Olson, Randal S., Nathan Bartley, Ryan J. Urbanowicz, and Jason H. Moore. 2016. "Evaluation of a Tree-Based Pipeline Optimization Tool for Automating Data Science." In *Proceedings of the Genetic and Evolutionary Computation Conference 2016* (GECCO '16), 485–492. Denver, CO: Association for Computing Machinery. https://doi.org/10.1145/2908812.2908918.

O'Neil, Cathy. 2016. *Weapons of Math Destruction: How Big Data Increases Inequality and Threatens Democracy.* New York: Crown.

Osborne, Jason. 2008. *Best Practices in Quantitative Methods.* Thousand Oaks, CA: SAGE Publications.

Oxford Internet Institute. 2020. "OII MSc in Social Data Science." https://www.oii.ox.ac.uk/study/msc-in-social-data-science/.

Pakhomov, Serguei V., Steven J. Jacobsen, Christopher G. Chute, and Veronique L. Roger. 2008. "Agreement between Patient-Reported Symptoms and Their Documentation in the Medical Record." *American Journal of Managed Care* 14 (8): 530–539.

Palen, Leysia, and Kenneth M. Anderson. 2016. "Crisis Informatics—New Data for Extraordinary Times." *Science* 353 (6296): 224–225. https://doi.org/10.1126/science.aag2579.

Passi, Samir, and Steven Jackson. 2017. "Data Vision: Learning to See through Algorithmic Abstraction." In *Proceedings of the 2017 ACM Conference on Computer Supported Cooperative Work and Social Computing* (CSCW '17), 2436–2447. Portland, OR: Association for Computing Machinery. https://doi.org/10.1145/2998181.2998331.

Passi, Samir, and Steven J. Jackson. 2018. "Trust in Data Science: Collaboration, Translation, and Accountability in Corporate Data Science Projects." *Proceedings of the ACM on Human–Computer Interaction* 2 (CSCW): 1–28. https://doi.org/10.1145/3274405.

Paul, Michael J. 2012. "Mixed Membership Markov Models for Unsupervised Conversation Modeling." In *Proceedings of the 2012 Joint Conference on Empirical Methods in Natural Language Processing and Computational Natural Language Learning* (EMNLP-CoNLL '12), 94–104. Jeju Island, Korea: Association for Computational Linguistics.

Pauwels, Luc, ed. 2006. *Visual Cultures of Science: Rethinking Representational Practices in Knowledge Building and Science Communication.* Hanover, NH: Dartmouth College Press.

Pérez, Fernando, and Brian E. Granger. 2007. "IPython: A System for Interactive Scientific Computing." *Computing in Science Engineering* 9 (3): 21–29. https://doi.org/10.1109/MCSE.2007.53.

Pilkonis, Paul A., Lan Yu, Nathan E. Dodds, Kelly L. Johnston, Catherine C. Maihoefer, and Suzanne M. Lawrence. 2014. "Validation of the Depression Item Bank from the Patient-Reported Outcomes Measurement Information System (PROMIS®) in a Three-Month Observational Study." *Journal of Psychiatric Research* 56 (September): 112–119. https://doi.org/10.1016/j.jpsychires.2014.05.010.

Pine, Kathleen H., and Max Liboiron. 2015. "The Politics of Measurement and Action." In *Proceedings of the 33rd Annual ACM Conference on Human Factors in Computing Systems* (CHI '15), 3147–3156. Seoul: Association for Computing Machinery. https://doi.org/10.1145/2702123.2702298.

Poon, Sarah S., Rollin C. Thomas, Cecilia R. Aragon, and Brian Lee. 2008. "Context-Linked Virtual Assistants for Distributed Teams: An Astrophysics Case Study." In *Proceedings of the 2008 ACM Conference on Computer Supported Cooperative Work* (CSCW '08), 361–370. San Diego, CA: Association for Computing Machinery. https://doi.org/10.1145/1460563.1460623.

Quinton, Sarah, and Nina Reynolds. 2018. *Understanding Research in the Digital Age.* London: SAGE Publications.

Rattenbury, Tye, Joseph M. Hellerstein, Jeffrey Heer, Sean Kandel, and Connor Carreras. 2017. *Principles of Data Wrangling: Practical Techniques for Data Preparation.* Sebastopol, CA: O'Reilly Media.

Reason, Peter, and Hilary Bradbury, eds. 2005. *Handbook of Action Research: Concise Paperback Edition.* London: SAGE Publications.

Reinecke, Katharina, Minh Khoa Nguyen, Abraham Bernstein, Michael Näf, and Krzysztof Z. Gajos. 2013. "Doodle around the World: Online Scheduling Behavior Reflects Cultural Differences in Time Perception and Group Decision-Making." In *Proceedings of the 2013 Conference on Computer Supported Cooperative Work* (CSCW'13), 45–54. San Antonio, TX: Association for Computing Machinery.

Rich, Adrienne. 1986. *Your Native Land, Your Life: Poems.* New York: Norton.

Rokem, Ariel, Cecilia Aragon, Anthony Arendt, Brittany Fiore-Gartland, Bryna Hazelton, Joseph Hellerstein, Bernease Herman, Bill Howe, Ed Lazowska, Micaela Parker et al. 2015. "Building an Urban Data Science Summer Program at the University of Washington Escience Institute." *Bloomberg Data for Good Exchange Conference.* https://www.bloomberg.com/lp/d4gx-2015/.

Rovatsos, Michael, Brent Mittelstadt, and Ansgar Koene. 2019. "Landscape Summary: Bias in Algorithmic Decision-Making: What Is Bias in Algorithmic Decision-Making, How Can We Identify It, and How Can We Mitigate It?" Centre for Data Ethics in Innovation. https://doi.org/10.17639/6h14-5t34.

Rule, Adam, Aurélien Tabard, and James D. Hollan. 2018. "Exploration and Explanation in Computational Notebooks." In *Proceedings of the 2018 CHI Conference on Human Factors in Computing Systems* (CHI '18), 32. Montreal: Association for Computing Machinery.

Rule, Adam, Ian Drosos, Aurélien Tabard, and James D Hollan. 2018. "Aiding Collaborative Reuse of Computational Notebooks with Annotated Cell Folding." In *Proceedings of the ACM on Human–Computer Interaction* 2 (CSCW): 150.

Saleiro, Pedro, Benedict Kuester, Loren Hinkson, Jesse London, Abby Stevens, Ari Anisfeld, Kit T. Rodolfa, and Rayid Ghani. 2019. "Aequitas: A Bias and Fairness Audit Toolkit." *ArXiv:1811.05577 [Cs.LG]*, April. http://arxiv.org/abs/1811.05577.

Saxena, Devansh, Karla Badillo-Urquiola, Pamela J. Wisniewski, and Shion Guha. 2020. "A Human-Centered Review of Algorithms Used within the US Child Welfare System." In *Proceedings of the 2020 CHI Conference on Human Factors in Computing Systems* (CHI '20), 1–15. Honolulu: Association for Computing Machinery. https://doi.org/10.1145/3313831.3376229.

Scharfenberg, David. 2018. "Computers Can Solve Your Problem. You May Not Like the Answer." *Boston Globe*, September 21. https://apps.bostonglobe.com/ideas/graphics/2018/09/equity-machine.

Schön, Donald A. 1984. *The Reflective Practitioner: How Professionals Think in Action.* New York: Basic.

Schön, Donald A. 2002. "From Technical Rationality to Reflection-in-Action." In *Supporting Lifelong Learning: Perspectives on Learning*, edited by Ann Hanson, 40–61. London: Routledge.

Schwartz, Becca, and Gina Neff. 2019. "The Gendered Affordances of Craigslist 'New-in-Town Girls Wanted' Ads." *New Media & Society* 21 (11–12): 2404–2421. https://doi.org/10.1177/1461444819849897.

Scott, James C. 2020. *Seeing Like a State: How Certain Schemes to Improve the Human Condition Have Failed.* New Haven, CT: Yale University Press.

Seaver, Nick. 2015. "The Nice Thing about Context Is That Everyone Has It." *Media, Culture & Society* 37 (7): 1101–1109. https://doi.org/10.1177/0163443715594102.

Seaver, Nick. 2017. "Algorithms as Culture: Some Tactics for the Ethnography of Algorithmic Systems." *Big Data & Society* 4 (2): 1–12. 205395171773810. https://doi.org/10.1177/2053951717738104.

Segel, Edward, and Jeffrey Heer. 2010. "Narrative Visualization: Telling Stories with Data." *IEEE Transactions on Visualization and Computer Graphics* 16 (6): 1139–1148. https://doi.org/10.1109/TVCG.2010.179.

Selbst, Andrew D., danah boyd, Sorelle A. Friedler, Suresh Venkatasubramanian, and Janet Vertesi. 2019. "Fairness and Abstraction in Sociotechnical Systems." In *Proceedings of the Conference on Fairness, Accountability, and Transparency* (FAT* '19), 59–68. New York: Association for Computing Machinery. https://doi.org/10.1145/3287560.3287598.

Shilton, Katie 2013. "Values Levers: Building Ethics into Design." *Science, Technology & Human Values* 38 (3): 374–397.

Shin, Donghee, and Yong Jin Park. 2019. "Role of Fairness, Accountability, and Transparency in Algorithmic Affordance." *Computers in Human Behavior* 98 (September): 277–284. https://doi.org/10.1016/j.chb.2019.04.019.

Siodmonk, Andrea. 2020. "Lab Long Read: Human-Centred Policy? Blending 'Big Data' and 'Thick Data' in National Policy." UK Policy Lab. https://openpolicy.blog.gov.uk/2020/01/17/lab-long-read-human-centred -policy-blending-big-data-and-thick-data-in-national-policy/.

Smith, Linda Tuhiwai. 2013. *Decolonizing Methodologies: Research and Indigenous Peoples.* London: Zed Books.

Snyder, Jaime. 2017. "Vernacular Visualization Practices in a Citizen Science Project." In *Proceedings of the 2017 ACM Conference on Computer Supported Cooperative Work and Social Computing* (CSCW '17), 2097–2111. Portland, OR: Association for Computing Machinery. https://doi.org/10.1145/2998181.2998239.

Snyder, Jaime, and Katie Shilton. 2019. "Spanning the Boundaries of Data Visualization Work: An Exploration of Functional Affordances and Disciplinary Values." In *iConference 2019: Information in Contemporary Society* (Lecture Notes in Computer Science), 63–75. Cham, Switzerland: Springer.

Snyder, Jaime, Elizabeth Murnane, Caitie Lustig, and Stephen Voida. 2019. "Visually Encoding the Lived Experience of Bipolar Disorder." In *Proceedings of the 2019 CHI Conference on Human Factors in Computing Systems* (CHI '19), 1–14. Glasgow: Association for Computing Machinery. https://doi.org/10.1145/3290605 .3300363.

Snyder, Jaime, Katie Shilton, and Sara Anderson. 2016. "Observing the Materiality of Values in Information Systems Research." In *Proceedings of the Hawaii International Conference on System Sciences (HICSS)*. Los Alamitos: IEEE.

Sorell, Tom, and Heather Draper. 2014. "Robot Carers, Ethics, and Older People." *Ethics and Information Technology* 16 (3): 183–195. https://doi.org/10.1007/s10676-014-9344-7.

Star, Susan Leigh. 2002. "Got Infrastructure? How Standards, Categories, and Other Aspects of Infrastructure Influence Communication." In *2nd Social Study of IT Workshop at the LSE ICT and Globalization.*

Star, Susan Leigh, and Anselm Strauss. 1999. "Layers of Silence, Arenas of Voice: The Ecology of Visible and Invisible Work." *Computer Supported Cooperative Work (CSCW)* 8 (1–2): 9–30.

Stein, Manuel, Halldór Janetzko, Daniel Seebacher, Alexander Jäger, Manuel Nagel, Jürgen Hölsch, Sven Kosub, Tobias Schreck, Daniel Keim, and Michael Grossniklaus. 2017. "How to Make Sense of Team Sport Data: From Acquisition to Data Modeling and Research Aspects." *Data* 2 (1): 2. https://doi.org/10.3390/data2010002.

Stoyanovich, Julia, Bill Howe, Serge Abiteboul, Gerome Miklau, Arnaud Sahuguet, and Gerhard Weikum. 2017. "Fides: Towards a Platform for Responsible Data Science." In *Proceedings of the 29th International Conference on Scientific and Statistical Database Management* (SSDBM '17), 1–6. Chicago: Association for Computing Machinery.

Strasser, Bruno J. 2006. "Collecting and Experimenting: The Moral Economies of Biological Research, 1960s–1980s." *Preprints of the Max-Planck Institute for the History of Science*, 310: 105–123.

Strohmayer, Angelika, Jenn Clamen, and Mary Laing. 2019. "Technologies for Social Justice: Lessons from Sex Workers on the Front Lines." In *Proceedings of the 2019 CHI Conference on Human Factors in Computing Systems* (CHI '19), 1–14. Glasgow: Association for Computing Machinery. https://doi.org/10.1145/3290605 .3300882.

Sweeney, Latanya. 2000. "Simple Demographics Often Identify People Uniquely." *Health (San Francisco)* 671: 1–34.

Tabard, Aurélien, Wendy E. Mackay, and Evelyn Eastmond. 2008. "From Individual to Collaborative: The Evolution of Prism, a Hybrid Laboratory Notebook." In *Proceedings of the ACM 2008 Conference on Computer Supported Cooperative Work* (CSCW '08), 569. San Diego, CA: ACM Press. https://doi.org/10.1145 /1460563.1460653.

Tanweer, Anissa. 2018. "Data Science of the Social: How the Practice Is Responding to Ethical Crisis and Spreading across Sectors." PhD diss., University of Washington. https://digital.lib.washington.edu:443/researchworks /handle/1773/43343.

Tanweer, Anissa, Brittany Fiore-Gartland, and Cecilia Aragon. 2016. "Impediment to Insight to Innovation: Understanding Data Assemblages through the Breakdown–Repair Process." *Information, Communication & Society* 19 (6): 736–752. https://doi.org/10.1080/1369118X.2016.1153125.

Tanweer, Anissa, Margaret Drouhard, Brittany Fiore-Gartland, Nicholas Bolten, Jess Hamilton, Kaicheng Tan, and Anat Caspi. 2017. "Mapping for Accessibility: A Case Study of Ethics in Data Science for Social Good." *Bloomberg Data for Good Exchange Conference*. September 24, New York, NY.

Taylor, Alex S., Siân Lindley, Tim Regan, David Sweeney, Vasillis Vlachokyriakos, Lillie Grainger, and Jessica Lingel. 2015. "Data-in-Place: Thinking through the Relations between Data and Community." In *Proceedings of the 33rd Annual ACM Conference on Human Factors in Computing Systems* (CHI '15), 2863–2872. Seoul: Association for Computing Machinery. https://doi.org/10.1145/2702123.2702558.

Taylor, Jennyfer L., Wujal Wujal Aboriginal Shire Council, Alessandro Soro, Paul Roe, and Margot Brereton. 2019, December. "A Relational Approach to Designing Social Technologies That Foster Use of the Kuku Yalanji Language." In *Proceedings of the 31st Australian Conference on Human–Computer-Interaction* (OZCHI '19), 161–172. Fremantle, Western Australia: Association for Computing Machinery.

Taylor, Linnet. 2017. "What Is Data Justice? The Case for Connecting Digital Rights and Freedoms Globally." *Big Data & Society* 4 (2): 1–14. https://doi.org/10.1177/2053951717736335.

Teal, Tracy K., Karen A. Cranston, Hilmar Lapp, Ethan White, Greg Wilson, Karthik Ram, and Aleksandra Pawlik. 2015. "Data Carpentry: Workshops to Increase Data Literacy for Researchers." *International Journal of Digital Curation* 10 (1): 135–143. https://doi.org/10.2218/ijdc.v10i1.351.

Terzi, Duygu Sinanc, Ramazan Terzi, and Seref Sagiroglu. 2015. "A Survey on Security and Privacy Issues in Big Data." In *2015 10th International Conference for Internet Technology and Secured Transactions (ICITST)*, 202–207. IEEE.

Thomas, James J., and Kristin A. Cook, eds. 2005. *Illuminating the Path: Research and Development Agenda for Visual Analytics*. National Visualization and Analytics Center and IEEE Press.

Thomas, Suzanne L., Dawn Nafus, and Jamie Sherman. 2018. "Algorithms as Fetish: Faith and Possibility in Algorithmic Work." *Big Data & Society* 5 (1): 205395171775155. https://doi.org/10.1177/2053951717751552.

Thompson, Stuart A., and Charlie Warzel. 2019. "Twelve Million Phones, One Dataset, Zero Privacy." *New York Times*, December 19. https://www.nytimes.com/interactive/2019/12/19/opinion/location-tracking-cell-phone.html.

Treviranus, Jutta. 2014. "Leveraging the Web as a Platform for Economic Inclusion." *Behavioral Sciences & the Law* 32 (1): 94–103. https://doi.org/10.1002/bsl.2105.

Tufekci, Zeynep. 2014. "Big Questions for Social Media Big Data: Representativeness, Validity and Other Methodological Pitfalls." In *Proceedings of the Eighth International AAAI Conference on Weblogs and Social Media* (ICWSM), 505–514. Ann Arbor, MI: Association for the Advancement of Artificial Intelligence.

Tufte, Edward R. 1983. *The Visual Display of Quantitative Information*. Cheshire, CT: Graphics Press.

Tufte, Edward R. 1990. *Envisioning Information*. Cheshire, CT: Graphics Press.

UK Government Digital Service. 2020. "Data Ethics Framework." https://assets.publishing.service.gov.uk/government/uploads/system/uploads/attachment_data/file/923108/Data_Ethics_Framework_2020.pdf.

Valentino-DeVries, Jennifer, Natasha Singer, Michael H. Keller, and Aaron Krolik. 2018. "Your Apps Know Where You Were Last Night, and They're Not Keeping It Secret." *New York Times*, December 10.

Vedres, Balazs, and Orsolya Vasarhelyi. 2019. "Gendered Behavior as a Disadvantage in Open Source Software Development." *EPJ Data Science* 8 (1): 25. https://doi.org/10.1140/epjds/s13688-019-0202-z.

Velden, Theresa, Asif-ul Haque, and Carl Lagoze. 2010. "A New Approach to Analyzing Patterns of Collaboration in Co-Authorship Networks: Mesoscopic Analysis and Interpretation." *Scientometrics* 85 (1): 219–242. https://doi.org/10.1007/s11192-010-0224-6.

Vella, Kellie, Jessica L. Oliver, Dema Tshering, Margot Brereton, and Paul Roe. 2020. "Ecology Meets Computer Science: Designing Tools to Reconcile People, Data, and Practices." In *Proceedings of the 2020 CHI Conference on Human Factors in Computing Systems* (CHI '20), 1–13. Honolulu: Association for Computing Machinery. https://doi.org/10.1145/3313831.3376663.

Vertesi, Janet. 2015. *Seeing like a Rover: How Robots, Teams, and Images Craft Knowledge of Mars*. Chicago: University of Chicago Press.

Voigt, Paul, and Axel Von dem Bussche. 2017. *The EU General Data Protection Regulation (GDPR): A Practical Guide*. Cham, Switzerland: Springer International.

Wang, April Yi, Anant Mittal, Christopher Brooks, and Steve Oney. 2019. "How Data Scientists Use Computational Notebooks for Real-Time Collaboration." *Proceedings of the ACM on Human–Computer Interaction* 3 (CSCW): 39.

Wang, Dakuo, Justin D. Weisz, Michael Muller, Parikshit Ram, Werner Geyer, Casey Dugan, Yla Tausczik, Horst Samulowitz, and Alexander Gray. 2019. "Human-AI Collaboration in Data Science: Exploring Data Scientists' Perceptions of Automated AI." *Proceedings of the ACM on Human–Computer Interaction* 3 (CSCW): 211.

Wang, Dakuo, Parikshit Ram, Daniel Karl I. Weidele, Sijia Liu, Michael Muller, Justin D. Weisz, Abel N. Valente, Arunima Chaudhary, Dustin Torres, Horst Samulowitz, and L. Amini. 2020. "AutoAI: Automating the End-to-End AI Lifecycle with Humans-in-the-Loop." In *Proceedings of the 25th International Conference on Intelligent User Interfaces Companion* (IUI '20), 77–78. Cagliari, Italy: Association for Computing Machinery. https://doi.org/10.1145/3379336.3381474.

Wang, Tricia. 2016. "Why Big Data Needs Thick Data." Medium, December 5. https://medium.com/ethnography-matters/why-big-data-needs-thick-data-b4b3e75e3d7.

Ware, Colin. 2020. *Information Visualization: Perception for Design*. 4th ed. Cambridge, MA: Morgan Kauffman. https://www.sciencedirect.com/science/book/9780128128756.

Wasserman, Stanley, and Katherine Faust. 1994. *Social Network Analysis: Methods and Applications*. Vol. 8. Cambridge: Cambridge University Press. https://doi.org/10.1017/CBO9780511815478.

Wible, Pamela. 2016. "When Someone Says Doctor, Do You Think Old White Guy? Tell the Truth." *Idealmedicalcare.org*, October 14. https://www.idealmedicalcare.org/when-someone-says-doctor-do-you-think-old-white-guy/.

Wickham, Hadley. 2014. "Tidy Data." *Journal of Statistical Software* 59 (10): 1–23.

Wickham, Hadley, and Garrett Grolemund. 2016. *R for Data Science: Import, Tidy, Transform, Visualize, and Model Data*. Sebastopol, CA: O'Reilly Media.

Wiggins, Chris. 2019. "Data Science at the *New York Times*." Presented at Rev 2: Data Science Leaders Summit, New York, May 24. https://blog.dominodatalab.com/data-science-at-the-new-york-times/.

Wilcox, David. 2004. "A Short Guide to Partnerships." Partnerships Online. Accessed June 2, 2020. http://www.partnerships.org.uk/part/index.htm.

Williams, Marian G., and Viv Begg. 1993. Translation between Software Designers and Users. *Communications of the ACM* 36 (6): 102–103.

Williams, Rua M., and LouAnne E. Boyd. 2019. "Prefigurative Politics and Passionate Witnessing." In *The 21st International ACM SIGACCESS Conference on Computers and Accessibility* (ASSETS '19), 262–266. New York: Association for Computing Machinery. https://doi.org/10.1145/3308561.3355617.

Wilson, Greg, Jennifer Bryan, Karen Cranston, Justin Kitzes, Lex Nederbragt, and Tracy K. Teal. 2017. "Good Enough Practices in Scientific Computing." Edited by Francis Ouellette. *PLOS Computational Biology* 13 (6): e1005510. https://doi.org/10.1371/journal.pcbi.1005510.

Wilson, Timothy D. 2004. *Strangers to Ourselves*. Cambridge, MA: Harvard University Press.

Winner, Langdon. 1978. *Autonomous Technology: Technics-out-of-Control as a Theme in Political Thought*. Cambridge, MA: MIT Press.

Woelfer, Jill Palzkill, Amy Iverson, David G. Hendry, Batya Friedman, and Brian T. Gill. 2011. "Improving the Safety of Homeless Young People with Mobile Phones: Values, Form, and Function." In *Proceedings of the SIGCHI Conference on Human Factors in Computing Systems* (CHI '11), 1707–1716. Vancouver, BC: Association for Computing Machinery.

Wuchty, Stefan, Benjamin F. Jones, and Brian Uzzi. 2007. "The Increasing Dominance of Teams in Production of Knowledge." *Science* 316 (5827): 1036–1039. https://doi.org/10.1126/science.1136099.

Wulf, William A. 1993. "The Collaborative Opportunity." *Science* 261 (5123): 854–855. https://doi.org/10.1126/science.8346438.

Xing, Wanli, Rui Guo, Eva Petakovic, and Sean Goggins. 2015. "Participation-Based Student Final Performance Prediction Model through Interpretable Genetic Programming: Integrating Learning Analytics, Educational Data Mining and Theory." *Computers in Human Behavior* 47: 168–181.

Ye, Nong. 2013. *Data Mining: Theories, Algorithms, and Examples*. Boca Raton, FL: CRC Press.

Yu, Shui. 2016. "Big Privacy: Challenges and Opportunities of Privacy Study in the Age of Big Data." *IEEE Access* 4: 2751–2763.

Zegura, Ellen, Carl DiSalvo, and Amanda Meng. 2018. "Care and the Practice of Data Science for Social Good." In *Proceedings of the 1st ACM SIGCAS Conference on Computing and Sustainable Societies* (COMPASS '18), 1–9. San Jose, CA: Association for Computing Machinery. https://doi.org/10.1145/3209811.3209877.

Zhang, Amy X., Michael Muller, and Dakuo Wang. 2020. "How Do Data Science Workers Collaborate? Roles, Workflows, and Tools." *Proceedings of the ACM on Human–Computer Interaction* 4 (CSCW1): 1–23.

Zheng, Kai, David A. Hanauer, Nadir Weibel, and Zia Agha. 2015. "Computational Ethnography: Automated and Unobtrusive Means for Collecting Data in Situ for Human–Computer Interaction Evaluation Studies." In *Cognitive Informatics for Biomedicine*, edited by V. Patel, T. Kannampallil, and D. Kaufman, 111–140. Cham, Switzerland: Springer International. https://doi.org/10.1007/978-3-319-17272-9_6.

Zhu, Xiaojin, and Andrew B. Goldberg. 2009. *Introduction to Semi-Supervised Learning*. San Rafael, CA: Morgan & Claypool. https://doi.org/10.2200/S00196ED1V01Y200906AIM006.

Zuboff, Shoshana. 2019. *The Age of Surveillance Capitalism: The Fight for a Human Future at the New Frontier of Power*. New York: Public Affairs.

Index